CONFRONTING CATASTROPHE

New perspectives on natural disasters

David Alexander
University of Massachusetts

TERRA

© David E. Alexander 2000

First published in 2000 by Terra Publishing

Terra Publishing
PO Box 315, Harpenden, Hertfordshire AL5 2ZD, England
Telephone: +44 (0)1582 762413
Fax: +44 (0)870 055 8105
Website: www.rjpc.demon.co.uk
E-mail: publishing@rjpc.demon.co.uk

ISBNs: 1-903544-00-9 hardback
           1-903544-01-7 paperback

10 09 08 07 06 05 04 03 02 01 00
10 9 8 7 6 5 4 3 2 1

British Library Cataloguing-in-Publication Data
A CIP record for this book is available from the British Library

Library of Congress Cataloging-in-Publication Data are available

Typeset in Palatino and Helvetica
Printed and bound by Biddles Limited, Guildford and King's Lynn, England

Contents

CONTENTS

iv

Preface

Despite 75 years of constantly accelerating progress in the field, there is something profoundly unsatisfying about modern studies of natural disaster. Theories, and the practice that has generated them, have not fully explained why the toll of deaths and losses is rising as steeply as it is (Alexander 1993: 2). I believe this is because disaster has not been considered sufficiently in light of present-day worldwide trends and tendencies. In the way that it illuminates a subtle interplay of progress and setback, catastrophe opens a window upon the inner workings of society. To make full use of this insight we must broaden the context of our studies and view disasters in terms of how the world is changing. This requires some analysis of history and some of human cultures. Although "culture" is an elusive concept, and one that is open to myriad different definitions, it is nevertheless fundamental to any understanding of the impact of disasters, for it embodies the historical imprint of events, as these are carried forward into the future through the mediation of experience and knowledge.

Disasters are holistic phenomena. I believe that they should not be studied in an *interdisciplinary* way, which implies obstacles to be surmounted, but through a *non-disciplinary* approach, which suggests a lack of boundaries. The nature of the problem to be solved should determine the methods applied to it, and the key to success is to discover and bring to light the links between disparate phenomena and events. Put more simply, we need to achieve a better marriage between the physical and social sides of disaster studies. So far, there has been a remarkable failure of nerve in this respect. Academics have spent too much time defending their own territories and propounding incomplete theories. At times one wonders whether we really seek to relieve human suffering by virtue of our erudition or merely attempt to make intellectual capital out of the sum of human misery. Nevertheless, during the 1990s, the International Decade for Natural Disaster Reduction, studies of calamity have achieved increasing respectability and have now accumulated a substantial body of knowledge and theories. But they have not produced an adequate unified theory to explain the phenomena under study according to simple universal principles. Perhaps this is an impossible goal, although it remains an alluring one: the "philosopher's stone" of the nascent discipline of "disasterology".

One reason for the failure to unify is the increasing divergence between the developed and developing worlds, and obviously not merely in terms of the strongly differential impact of disasters. The industrialized countries are rich in technological capital, whereas the developing nations abound in human resources. There is an increasing sense of distance between the two. As a result, the few cultural bridges that might facilitate the transfer of technology for mitigation are being gradually destroyed by exploitation, forced militarization, economic manipulation and widening income gaps. Except where there is a strong sense of kinship, there tends to be an unwillingness to learn from other societies. Yet many fundamental principles are universal: for example, disaster mitigation must go hand in hand with socio-economic development in any society, rich or poor. Equity concerns are ubiquitous, even though the proposed solutions tend not to be.

My intention in this book is not to produce a systematic account of disasters, or another distillation of the current literature, but to give the reader a personal view. It is all too easily forgotten that disaster studies should be a marriage of learning and experience, the latter as practical as possible. Too much abstraction is dangerous – it obfuscates rather than illuminates the situation. My aim here is to confront some of the awkward realities of disaster as directly as possible without resorting to the dry, abstract debate that characterizes so many learned discourses on the subject of catastrophe, or to the miserable trail of unconnected anecdotes that constitutes the main alternative approach. My solution is to use an alternation of principles and examples, using the latter as occasion to comment on what general factors they illuminate. If this bares some prejudices and predilections, then so be it: *dulce est desipere in loco.*[1]

I would like to thank Lorna Stinchfield, Carol Vogel and Annabelle Lucas for typing the first draft of this book from my notes, and Roland Sarti for reviewing part of the text.

I dedicate this book to Joyce M. Alexander and Eric E. Alexander on the occasion of their fiftieth wedding anniversary.

David Alexander
San Casciano in Val di Pesa, Italy
June 2000

---

1. "It is fun to mess around a bit, when there is enough time to do so" (Horace, *Odes*).

CHAPTER 1

Introduction

μανὲ ὅτι αἱ συμφοραὶ ἶ ὦν ἀνϑρώϋων ᾽άρχονσικαὶ
οὐκὶ ὦνϑρωποι τῶν σνμφορέων μανὲ ὅτί
You must learn that the disasters
govern the man, not man the disasters

A few days before I sat down to write this book and a few tens of kilometres from my desk, it began to rain heavily in the Versilian Mountains of northwest Italy. Some 415 mm of precipitation fell on the steep wooded slopes in only 24 hours; half a year's rainfall in two days. Suddenly, the mountain streams rose and transformed themselves into heaving torrents of mud and rock. They roared through the ancient stone villages that nestle in the narrow wooded valleys: Stazzema, Cardoso, Fornovalasco. The mud and water rose nimbly out of the beds of the torrents and flooded into the streets, the squares and the huddled groups of stone houses. Torn from its moorings, a propane tank danced in the current, spurting twin jets of gas into the rain-soaked air. Cars parked in a piazza rose in the flow, bobbed around for a while and one by one set off sedately down the valley to be smashed to pieces on the rocks below. An ancient stone bridge collapsed with a crash, and the rubble streamed away piece by piece in the midst of the hungry black water. Spates of mud and boulders rolled turbulently through the narrow streets, smashing vehicles and fixtures and demolishing the walls of houses. Thirteen people died in the floods. The boiling water took a four-year-old girl from her collapsing house, transported her body 20 km down stream and laid it to rest on the shore of the Tyrrhenian Sea, hideously disfigured. The narrow, winding roads through the valleys disappeared in a welter of landslides and meander scars. A kilometre of the main Tyrrhenian coast railway was buried under tonnes of mud, rocks and uprooted tree trunks; its bridges were washed away, leaving the track suspended over the cavernous holes left by the streams.

In short, a flash flood occurred with a recurrence interval of somewhere between 100 and 700 years, which was, for the area, a truly exceptional event.

1

INTRODUCTION

The full apparatus of civil protection rose to the occasion. Within hours the sky was thick with helicopters and the access roads were thronging with trucks and ambulances. Soldiers, volunteers, firemen, policemen, cabinet ministers, regional politicians, camera crews, even coach parties of tourists, converged on the stricken villages.[1] The people who live in the mountains are taciturn and accustomed to hardship, but many appeared dazed and incredulous as the men in orange jackets led them to safety.[2] Several hundred of them had lost their homes, and US$30 million of damage had been done by the floods.

In a world context the Versilian floods were a truly unexceptional event. They coincided with disastrous wildfires in Mongolia, catastrophic inundations in Yemen, Hurricane Arthur in Mexico, and several other extreme events. But in the minds and lives of everyone who experienced and survived them, the Italian floods will form a permanent marker, a point of reference with which to position other events in time and with which to measure their significance.[3] They were, after all, the worst floods in Tuscany since 3 November 1966, when the River Arno burst its banks in Florence and killed 31 of that city's inhabitants (Di Leva 1996).

Thus, in terms of the perception of those who participate in it, a disaster is a unique event. If time is linear, as the mediaeval mind would have it, and not cyclical, as Plato and Aristotle supposed it was, then disasters cannot repeat themselves (Gould 1987). Each time one occurs, the ingredients, the controlling parameters and the outcome variables are present in unique mixtures. But disasters are also subject to generalization. The common elements are nearly always present in terms of a well mapped spatial and temporal unfolding of more or less consecutive phases (see Figs 1.1, 1.2).[4] Thus, the survivors of the Versilian floods formed part of a much larger and more widespread group: according to Red Cross data, each year 130 000 people are killed, 90 000 are injured and 140 million are affected by an average total of more than 200 natural disasters (IFRCRCS 1998).[5]

Much attention has been given to the question of what a disaster actually represents.[6] Given that there are at least six schools of thought on the subject, collectively representing more than 30 different disciplines (Alexander 1991a;

---

1. The classic work on convergence behaviour is by Fritz (1957). Scanlon (1992) reconsidered the subject in the light of more recent findings.
2. See Wallace (1956) for an early investigation of the psychological "disaster syndrome". See also Horowitz (1986).
3. See Erikson (1976) for a classic study of disaster as a permanent marker in the lives of individuals and communities.
4. But see the critique of these phases given by Neal (1997).
5. One reason why deaths outnumber injuries is the high mortality in droughts, which is not accompanied by a toll of injuries.
6. See, for example, Quarantelli (1995).

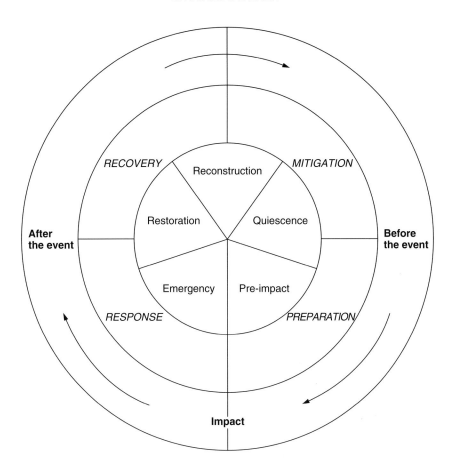

Figure 1.1 The disaster cycle.

1993: 12–14), the answer is bound to be complex and multifarious. In one sense, disaster is a window on society, a chance to observe the workings of social and cultural processes under an extreme duress that exposes their inner essence. Physically, one is dealing with destructive extreme events, socially with a phenomenon that puts human organization under stress or in crisis (Barton 1970, Gillespie 1988) and which puts human adaptability to the test. In economic terms, disasters result in the accelerated consumption of goods and services (Jones 1987); logistically they provide an opportunity to improve mitigation measures. Mitigation measures can be viewed as a "window of opportunity", in which a recent disaster sensitizes public and political opinion, and brings forth demands for improved safety in the future.[7]

7. See Solecki & Michaels (1994) for a practical study of this aspect of disasters.

(a) Developing country

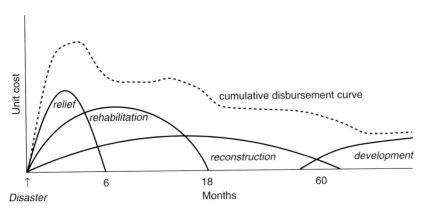

(b) Industrialized country

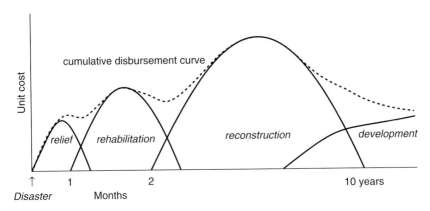

Figure 1.2 Stages in recovery from disaster: (a) for a developing country, (b) for an industrialized nation (modified from fig. 1 in Kirkby et al. 1997).

In order to examine these questions in detail, the book is divided into eight parts. To set the scene we begin by examining the question of how to define key terms and concepts in the study of natural disasters. We then proceed with an examination of the ways in which disaster has been studied. Although much progress has been made over the past 75 years, there have been many missed opportunities, and problems have arisen as a result of overspecialization and territoriality among academic disciplines. This section is followed by an inquiry into the relationship between disasters, society and culture, which provide one of the keys to the interpretation of human reactions to adverse events. Although culture is a fundamental variable, researchers have not given it full weight.

Following this, we examine the role of history in determining the present and, as far as can be ascertained, the future impact of natural disasters on human societies. The past is the key to many evolving trends in this field and it is therefore necessary to read the lessons of history attentively in order to understand what lies in store at this critical moment in human social evolution.

The fifth chapter will examine some of the more mechanistic aspects of adaptation to, and management of, disaster, with due consideration of the cultural and social underpinnings of the use of technology. Next is a chapter on some of the moral and philosophical issues that underpin the study and interpretation of natural catastrophe, including its relationship with other forms of disaster, such as those precipitated by armed conflict. This is especially pertinent to the penultimate section, which will examine the plight of developing countries affected by natural hazards, with the aid of examples intended to clarify the situation on the ground. The final section draws some of the threads together and presents a model that places natural disasters firmly within the context of change in modern society.

It is appropriate to begin with the vexed question of how to define natural disaster and its components.

CHAPTER 2

Definitions

This chapter will examine some contemporary interpretations of hazard and disaster and a few of the developing trends in analytical methods used by students of this field. This is intended to provide a basis for understanding the social, technological, moral and practical issues that combine to shape natural disasters as phenomena of the modern world.

The problem of how to define disaster[1] and its many components is far from solved and has been the subject of considerable debate in various disciplines.[2] Let us begin by examining the question of how to define four key terms: hazard, risk, vulnerability and disaster.

A **hazard** is an extreme geophysical event that is capable of causing a disaster. "Extreme" in this case signifies a substantial departure in either the positive or the negative direction from a mean or a trend: hence, flood disasters result from unusually high precipitation and river discharge, whereas drought disasters stem from unusually low values. The fundamental determinants of hazards are location, timing, magnitude and frequency. Many hazardous phenomena are recurrent in time and predictable in terms of location. For example, hurricanes (typhoons or intense tropical cyclones) occur between 5° and 25° north and south of the Equator and tend to be seasonal phenomena. In the North Atlantic Ocean they develop in late summer and autumn in response to the meteorological perturbations known as easterly waves (Hess & Elsner 1994, Diaz & Pulwarty 1997). For floods and earthquakes, magnitude of event corresponds to a given average recurrence interval. In fact, the **magnitude–frequency rule** states that

---

1. In this book, "disaster", "calamity" and "catastrophe" will be treated as synonyms. I do not believe there are any adequate grounds for defining these terms differently from one another.
2. See, for example, Bailey (1989), Kreps (1989), Kroll-Smith & Couch (1991) and Quarantelli (1998) among the sociologists, and Ball (1979), Frerks et al. (1995) and Kelly (1996) among the relief and development specialists. Further definitions are given in Alexander (1991a; 1993: ch. 1).

over a sufficient interval of time there will be many small events and few large ones (Wolman & Miller 1960). Hence, the average return period of small events is short and that of large events is long. In terms of the environmental changes wrought by geophysical events, one expects there to be a rough sort of equilibrium between the cumulative effects of small events and the occasional solitary impact of large events (Hewitt 1970; Smith 1996: 10–11).

It is perhaps unwise to be too glib about the magnitude–frequency rule. In the first place, detailed data are seldom available for long enough periods of time to allow magnitude–frequency graphs to be constructed with any degree of statistical certainty. To combat this problem, much statistical artifice is employed, but predictions tend to be unreliable when events of large magnitude and long return period are involved (this is especially true of volcanic eruptions, in which there may be no measurable periodicity at all). Environmental reconstruction and absolute dating have done much to fill in the gaps and reconstitute the time series of extreme events, but it is not uncommon to find that datable evidence is only partial: perhaps ancient flood deposits have been eroded away, or earthquakes have occurred without producing datable surface ruptures.[3] Secondly, in the timespan of human lives, average recurrence intervals can be highly irregular and hence difficult to predict. For hazards that have a meteorological origin, such as floods and droughts, actual recurrence intervals can be highly irregular and hence difficult to predict.[4] Thirdly, to some extent the size of disaster may be independent of the magnitude of the geophysical event. Thus, the mudflow that killed 144 people at Aberfan, South Wales, in 1966 involved only 75 000 m^3 of debris, which moved at walking pace through the schools and houses of the town. In contrast, no disaster was caused by the Sherman landslide of 1964, which sent 30 million m^3 of rock crashing down at more than 100 km/hr into an uninhabited valley of central Alaska (Alexander 1993: 9–10).

Lastly, in the Earth sciences there is an unresolved, and possibly unresolvable, debate about the significance of large infrequent events. In the physical landscape, does the cumulative effect of many small events outweigh the impact of the occasional cataclysm, or vice-versa (see Brunsden & Thornes 1979)? We may look upon the dilemma as a duel between uniformitarianism and catastrophism, a debate that has lasted for most of recorded history.[5] Among

---

3. Note that for the most part, regardless of whether geochronological methods are relative or absolute techniques, they can date only what is there, not what is missing because erosion has stripped it away, which must be inferred (Mahaney 1984).

4. Glantz (1982) and Mayer & Nash (1987) tackled some of the consequences associated with difficulties of prediction.

5. A classic reference here is the Frank Dawson Adams scholarly account of the history of the Earth sciences (Adams 1938). After 60 years it is still in print.

the Ancients, Aristotle, Plato, Strabo and Herodotus were uniformitarianists. Centuries later, Athanasius Kircher, John Ray, Nikolaus Steno and Abram Gottlob Werner were catastrophists. At the global scales, uniformitarianism underlay Wegener's theory of continental drift and much of the early formulation of plate tectonics. At the local scale, the uniformitarian approach was best exemplified by a paper in which M. Gordon Wolman and John P. Miller argued that the morphology of mid-latitude streams in humid temperate climates results from precipitation events and flood flows of modest dimensions and almost biennial occurrence (Wolman & Miller 1960). Lately, however, neocatastrophism has come back into fashion. Intense interest in mass extinctions has been coupled with the realization that global change can occur quite abruptly when certain turning points are reached. With the aid of a dose of millennialism (a form of prophecy to which scientists are by no means immune), the research community has been stimulated to look again at many phenomena in a neocatastrophist light (Dury 1980, Baker 1988). Even the uniformitarian view of small mid-latitude stream channels has been tempered by a re-evaluation of the significance of large flood deposits and deep erosional scars (Baker et al. 1988).

However, it is all too easy to oversimplify the uniformitarianism/catastrophism debate. In reality the two points of view need not – perhaps *should* not – be mutually exclusive. Leonardo da Vinci offered a good example of this: his studies of water flow revealed a considerable predilection for catastrophism, but his interpretations of sedimentation and stratification were firmly rooted in uniformitarianism (Alexander 1982a).

Whether it is the summation of relatively small events or the occasional occurrence of large cataclysms that causes disaster, it seems reasonable to suppose that there is a threshold value of the physical forces unleashed that defines the lower limit at which extreme geophysical phenomena are capable of causing disaster.[6] However, it is abundantly clear that such a value depends critically on the human impact of such forces: the vulnerability of people, society and the built environment may alone determine the magnitude at which an event becomes a disaster.[7] In this respect we define **natural hazards** as extreme events that originate in the biosphere, lithosphere, hydrosphere or atmosphere. The term is useful because it distinguishes such phenomena from **technological hazards** – including explosions, releases of toxic materials, episodes of severe contamination, structural collapses, and transportation, construction and manufacturing accidents – and from **social hazards**, such as crowd crushes,

---

6. See Figures 9.4 and 9.5 (pp. 232–233).
7. Hewitt (1983) went some way towards sustaining this view in his "radical critique" of causality in disaster. It seems to be gathering ground: according to Cannon (1994), the emphasis on the economic and political causes of disasters seems to have reached some sections of the public, and some politicians and aid workers (see also Hendrickson 1998).

riots and terrorist incidents. However, many would argue that "natural" hazard is a misleading term, as very little is natural about phenomena in which the danger results largely from human decision making, land use and socio-economic activities, in as much as these impinge upon the predictable domain of extreme natural events (Hewitt 1997: ch. 3).

In the traditional linear view of hazards, an extreme geophysical event leads to a concentrated physical expenditure of energy that acts upon the distributed elements of human vulnerability and risk, tempered by any structural or non-structural mitigation measures that may be in place, to create the net impact of disaster. It is implied that the physical hazard causes the human disaster. But if human vulnerability and propensity to take risks by transforming or manipulating the natural environment are regarded as paramount, then the direction of causality is reversed. The vulnerability of the human environment causes disaster through the medium of the violence of the physical environment (Hewitt 1983). Magnitude and frequency have little intrinsic meaning in such a context, unless they are tempered with some measure of human invulnerability and risk.[8]

Risk can be defined as the likelihood, or more formally the probability, that a particular level of loss will be sustained by a given series of elements as a result of a given level of hazard impact. The elements at risk consist of populations, communities, the built environment, the natural environment, economic activities and services, which are under threat of disaster in a given area. **Total risk** consists of the sum of predictable deaths, injuries, destruction, damage, disruption, and costs of repair and mitigation (UNDRO 1982). It can also be regarded as

$$\text{total risk} = (\Sigma \text{ elements at risk}) \times (\text{hazard} \times \text{vulnerability}).$$

Much has been written about the concept of risk,[9] to the extent that it has crystallized into a fully fledged field of study.[10] The field is divided into three distinct parts: risk perception and communication, risk estimation, and risk management (Burton & Pushchak 1984). In essence, the public do not perceive

---

8. Cannon (1994: 19) argued that the vulnerability of individuals and groups of people who inhabit a given natural, social and economic space is differentiated according to their position in society. It derives in a complex way mainly from class, gender and ethnicity, and in the way that these are expressed in socio-economic terms. Thus, it reflects the tripartite effect of economic resilience in the face of hazard impacts, medical and sanitary resilience (including the presence or absence of healthcare and preventive medicine), and degree of personal and social preparedness.

9. According to Cannon (1994: 14), natural phenomena offer both risks and opportunities, but their effect is modified by unequal exposure to the former and unequal access to the latter.

10. See works by Douglas & Wildavsky (1982), Wilson (1991), Waterstone (1992), Cutter (1994) and Newman & Strojan (1998).

risks accurately and in many instances will not react rationally to the risks they face; hence, the importance of estimating and communicating the types and levels of risk before any attempt can be made to apply scarce resources to their rational management (Kasperson & Stallen 1990). Although relatively few hard data exist on the subject, it is generally held that risk reduction by prior mitigation is cheaper than disaster relief as a result of unmitigated risk.[11] However, most structural mitigation measures, and perhaps also non-structural methods, involve diminishing returns when a threshold is reached that defines the point at which further investment in mitigation is not justified by the returns in damage and casualties avoided (Russell 1970). Nevertheless, the threshold is a moveable one in relation to the value – monetary or intrinsic – that society places on the items to be protected (Alexander 1995a). Such is the mixture of predilection and prejudice that no rational model exists to predict society's choices.

Natural hazard risk, then, is dynamic and complex. The urbanization of a floodplain, the deliberate flaunting of anti-seismic building codes, the rapid rise in populations of coastal towns that are susceptible to hurricanes, and the spread of precarious shanty towns on unstable tropical hillslopes – all represent ways in which risk increases. For example, although earthquake hazard is very moderate in Lagos, Nigeria (1995 population: 10.3 million), buildings tend to be vulnerable to seismic damage there, the population growth rate is 5.7 per cent per annum, and 23 per cent of Nigeria's urban population now live in the city. This tends to propel Lagos into a higher risk category than the hazard alone would suggest, and it represents a rather striking way in which the parameters of natural disaster can be altered over time (Kakhandiki & Shah 1998: 49). Gradual processes of organization, investment, training and increase in technical capacity represent ways in which civil protection can be used to reduce risk levels or at least provide a counterbalance to the processes of increase in risk.

However, we are still a long way from creating a culture of risk reduction, in which hazards are treated in a sober, responsible way and above all in the terms of **relative risk**: the need to reduce the risks that pose the greatest threats to society (Johnson & Covello 1987, Newman & Strojan 1998). Neither the data nor the economic means exist for full-scale risk reduction, even if the impetus were forthcoming from society, which it is not. Like the physical properties of strength and friction, risk comes into full being only when mobilized by forces, in this case the destructive power of hazards. For the rest of the time it remains an abstract concept, and the public are congenitally unable to view probabilities

---

11. See, for example, the figures quoted by Leighton (1976), which show consistently high benefit–cost ratios for prior mitigation of natural hazards, in comparison with damages avoided. In the late 1990s, bulletins issued by the US Federal Emergency Management Agency quoted a figure of US$2–3 saved for every dollar spent on valid mitigation measures (see http://www.fema.gov/).

rationally, hence the attraction of lotteries and other forms of gambling (Freudenburg 1988, Shrader-Frechette 1990).

If risk is one side of the coin, its other side is **vulnerability**, which we may loosely define as potential for losses or other adverse impacts. People, buildings, ecosystems or human activities threatened with disaster are vulnerable. The literature on risk and vulnerability is sometimes hazy about the distinction between the two concepts.[12] Essentially, vulnerability refers to the potential for casualty, destruction, damage, disruption or other form of loss with respect to a particular element. Risk combines this with the probable size of impact to be expected from a known magnitude of hazard. The latter can be considered as the manifestation of the agent that produces the loss. Building an unprotected factory next to a stream that is liable to flood creates both a situation of risk (probable flood damage) and an element of vulnerability (threatened property). Where vulnerability is not constantly present we may talk about the degree to which vulnerable elements are exposed to hazards: for example, a person who travels twice a day across a bridge that is vulnerable to collapse is exposed to this risk only for two brief periods in 24 hours. However, there are two reasons why the two concepts are often difficult to separate. First, the act of taking a risk creates a situation of vulnerability, whereas the existence of vulnerable elements in the light of known hazards poses a risk. Secondly, although the *presence* of vulnerability can usually be estimated without much knowledge of risk levels, it cannot be quantified without predicting the extent of damage and hence the strength of hazard, which is tantamount to estimating risk. Because of this it would be better if vulnerability were considered as one of the aspects of risk analysis, perhaps under a term such as **innate risk**. However, in reality, studies of vulnerability to hazards have tended to develop separately and in parallel to the burgeoning field of risk analysis.[13] Such is the tendency of academics to compartmentalize even the new interdisciplinary fields that they create.

Elements at risk of disaster have both an **intrinsic value** (their price at sale, or their estimated worth in intangible terms) and a **functional value** based on their role in sustaining the integrity and wellbeing of the community. One of the great paradoxes of the hazards field is that losses continue to rise despite the constant accumulation of knowledge on how to reduce risks and vulnerability (Berz 1994). In some cases there is undoubtedly a failure to utilize appropriate knowledge (Yin & Moore 1985); there may also be a failure to generate it by commissioning research. But for the most part it is clear that other mechanisms operate in society that inhibit the transformation of knowledge on hazards into

---

12. Many authors (e.g. Uitto 1998: 7) have confused vulnerability with exposure: in reality they are two complementary components of risk.
13. For example, compare Burton & Pushchak (1984) with Wisner (1993).

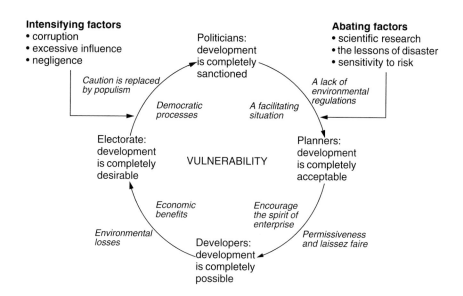

Intensifying factors
- corruption
- excessive influence
- negligence

Abating factors
- scientific research
- the lessons of disaster
- sensitivity to risk

Politicians: development is completely sanctioned

Caution is replaced by populism

Democratic processes

A facilitating situation

A lack of environmental regulations

Electorate: development is completely desirable

VULNERABILITY

Planners: development is completely acceptable

Economic benefits

Encourage the spirit of enterprise

Environmental losses

Developers: development is completely possible

Permissiveness and laissez faire

Figure 2.1   The vicious circle of increases in vulnerability: a process of positive feedback.

improved human safety. In the vicious circle of increases in vulnerability, the political, technological, social and cultural climates favour unprotected development and inhibit mitigation. Thus, politicians declare the development process to be fully acceptable, planners declare it to be fully permissible, developers declare it to be fully possible, and the electorate declare it to be fully desirable (Fig. 2.1). On the positive side, democracy is at work, enterprise is encouraged and economic benefits accrue. On the negative side, permissiveness and laxity are rampant, safeguards are absent, populism overrides prudence, and losses are inevitable. Corruption and excesses of influence and negligence dominate over the lessons of experience and research: vulnerability, risk and loss are strongly intertwined. The results of this are highly negative in a cyclical and recurrent way (Fig. 2.2). The vicious circle of increases in vulnerability represents a positive feedback situation in which bad policy and practice are self-generating, and development is, in the long-term, unsustainable. But positive feedback mechanisms are by nature transient: optimists may predict that mitigation will eventually triumph, pessimists that catastrophic destruction will be the final outcome. Sometimes, however, it requires a major disaster to stimulate wholesale improvements in safety. Indeed, it is common for the most significant pieces of legislation to be passed in the wake of large impacts.[14]

At any point in time, total vulnerability to hazards is governed by actions that increase risk levels, minus actions that mitigate them, but is tempered by factors of perception, which can be either positive or negative, depending on how

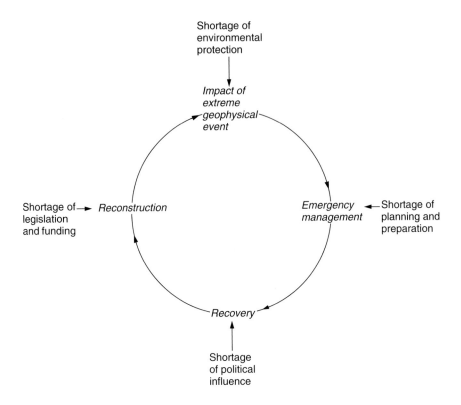

Figure 2.2 How the vicious circle of increasing vulnerability affects the disaster cycle.

hazards and risks are viewed and interpreted. Hazard and risk perception are subtly different,[15] but they have in common the fact that they can motivate either protective or risky responses to the threat of disaster and hence can contribute positively or negatively to vulnerability.

In an earlier work (Alexander 1991b) I developed the following analysis, which is based on simple conceptual equations and which I present here in modified form. In general terms, the impact of natural hazards is as follows:

$$B_N = B_T - D - C \tag{2.1}$$

---

14. For example, the Robert T. Stafford Disaster Relief and Emergency Assistance Act (PL93-288, as amended by PL100-707) was passed by the US Congress in 1993, shortly after Hurricane Andrew, the costliest disaster in US history at that time, had not only devastated a tract of southern Florida but also had called into question the very basis of Federal disaster assistance (Birkland 1997a).
15. For example, compare Otway & Thomas (1982) and Whyte (1986).

in which B_N is net benefit derived from occupying a natural hazard risk zone, B_T is the total benefit to be derived from occupying the area, D is the total loss caused in the area by natural hazards, and C is the total cost of adjustment to the hazard. Thus, natural hazard risks are a function of benefits and costs (Burton et al. 1993).

Looked at another way, total risk of death, injury, destruction, damage, homelessness, economic loss and amenity loss (R_T) is:

$$R_T = E \cdot R_S = E(H \cdot V) \tag{2.2}$$

where E is the elements at risk (the population, built environment, economic activities, and so on), $R_S = (H \cdot V)$ is specific risk, H is the natural hazard (the probability of impact in a given area over a given time period), and V is the vulnerability, expressed as the possible magnitude of losses (UNDRO 1982, Newman & Strojan 1998).

By combining equations (2.1) and (2.2) one obtains

$$B_N = B_T - kE\,(H \cdot V) - C \tag{2.3}$$

where $D = kE\,(H \cdot V)$, with the constant of proportionality $0 > k \leq 1$. Total losses averaged over time can be expressed as

$$D_T = E\,(H \cdot V) + C - B_T. \tag{2.4}$$

However, risk is also subject to stochastic factors, such that it is actually a function of exposure to hazard over time (t_E), vulnerability (V) of elements at risk (E), and the probability that a natural hazard will strike in a certain way (P):

$$R_T = \text{fcn}\,\{t_E,\, V(E),\, P\}. \tag{2.5}$$

Losses per unit time are thus

$$D_T = (t_E/t) \cdot (P_H/t) \cdot V(E). \tag{2.6}$$

In this formulation, vulnerability can be disaggregated as follows:

$$V = R_A - R_M \pm R_P, \tag{2.7}$$

in which R_A is the sum of factors that amplify risk (the result of bad practice), R_M is the sum of factors that mitigate risk (the result of good practice), and R_P is the risk perception factor, which can, on balance, be either negative or positive, depending on how accurately risks are viewed by those who sustain them.

15

The sum of elements at risk is therefore

$$\Sigma E \propto R_A - R_M \pm k_1 R_P, \qquad (2.8)$$

in which the constant k_1 can be positive or negative, depending on the effects of perceptual factors in either reducing or increasing risk levels through their influence on behaviour and thus on risk-taking propensities. The total costs of adjustment to natural hazards are proportional to the vulnerability of elements at risk, tempered by the degree of exposure of these to hazards and the probability that hazards will occur:

$$C \propto \{E(V), t_E, P(H)\} \propto R_T, \qquad (2.9)$$

Total risk is related to the prevailing climate of risk amplification or reduction:

$$R_T \propto \{E \cdot t_E \, [P(H) \cdot V] \cdot k_1 (R_A - R_M \pm R_P)\} \qquad (2.10)$$

that is, total risk is the product of the elements at risk, their specific vulnerability and the probability that a hazard will strike, within the prevailing climate of risk management, and given the level of exposure of the various elements. In other words, vulnerable subjects are exposed to hazards that may or may not strike over a given period of time, but, if they do, losses will be related to the changing pattern of vulnerability and mitigation.[16] In a highly developed society the risk of death by natural hazard is approximately 1 to 11 400 000 per person per year (Alexander 1993: 575). It is significantly higher in some of the world's worst-affected countries: for example, in China it is 1:375 000.

In practice, we can distinguish between several types and levels of vulnerability. To begin with, **positive vulnerability**, the "original sin" of hazard vulnerability, is determined solely by the fact that losses occur and are sustained. Lack of experience and lack of research mean that higher levels of knowledge are not attained. These would foster the sort of mitigation methods that reduce risks either by ameliorating the hazard or, more commonly, by lowering vulnerability.

Society effects a working consensus about what constitutes acceptable loss and hence what is an acceptable level of vulnerability. This may sound tautological if one regards no losses as truly acceptable, but, because absolute security can never be guaranteed, in practice some level of loss has to be tolerated. When losses become unacceptably high, the **threshold of loss tolerance** or

---

16. Similar analyses have been offered by Starr (1969), Okrent (1980), Covello & Mumpower (1985) and Vlasta (1996).

threshold of acceptable vulnerability (Burton et al. 1978) is crossed and a new consensus emerges about what level of vulnerability is acceptable. The threshold can vary in the long term directly with accumulated knowledge and demand for change, and in the short term inversely in proportion to the sum of antecedent losses; it is thus strongly linked to the **window of opportunity** (Lavell 1994, Solecki & Michaels 1994) for improvements in hazard mitigation.

The threshold of loss tolerance is determined by the level of knowledge of hazards and their impacts, which is proportional to the quantity of research on the subject and the degree to which it is utilized for mitigation. In a situation of **deprived vulnerability**, research results are neither diffused nor utilized sufficiently. Under **wilful vulnerability** they are deliberately ignored. Hence, in the latter case, the threshold of acceptable vulnerability is maintained artificially high by arbitrary factors such as corruption, negligence and politically inspired curtailment of the mitigation process (see Fig. 2.1).

When an area is first settled, **pristine vulnerability** prevails as a result of lack of experience of hazards. Thereafter, increases in regional and local urban development outstrip the production and utilization of research on hazards, which leads to deprived vulnerability. Once the development process has begun in earnest, there are social and economic pressures to maintain it. As noted above, these take the form of a self-reinforcing positive feedback process, in which growth begets more growth. The state of wilful vulnerability obtains if insufficient weight is given to hazard mitigation, if profit gives an incentive not to enact or enforce laws and norms, or if societal controls on individual actions are inadequate.

In this context it should be noted that much hazard research assumes a rosier picture of human attitudes than they merit.[17] It is assumed that the desire to do the right thing underlies human actions, however much these may contribute to the sum of vulnerability. Hence, there is practically no research on, for example, the role of organized crime in increasing vulnerability.[18] But this is hardly surprising, given the difficulties – indeed the risks – of obtaining good objective data on such a subject. Nevertheless, there are few data and even fewer models on the role of wantonness, corruption and exploitation in the creation of hazard losses. This makes it extremely difficult to assess whether the dark side of the human character plays a major or a minor role in disaster. For the time being, perhaps one should not assume, as so much of the research implicitly does, that human actions are essentially rational (albeit a rationality bounded by limitations of perception[19]), straightforward and honest.

---

17. Here, I part company with the sociologists, who tend to be more optimistic than I am (Dynes 1970, Quarantelli 1978, Dynes & Tierney 1994).
18. See Fearnside (1989) for an example of treatments of environmental despoliation in the light of factors that include politically inspired homicide. See also Olson (1995).

However, when disaster occurs on a massive scale, the full scope of human failings that have caused vulnerability to exist may be revealed in excruciating detail. That is what happened after the Izmit (Turkey) earthquake of 17 August 1999. This magnitude 7.4 event caused 15 500 deaths, 25 000 injuries and the collapse of more than 60 000 buildings. Patterns of damage were closely related to the quality of construction and the degree of observance of anti-seismic building practices, which, given the high number of casualties, can be judged to have been spectacularly low. Besides the **primary vulnerability** caused by the high susceptibility to catastrophic damage of so many buildings, **secondary vulnerability** was revealed in the poor response by the Turkish authorities to the initial emergency. Turkey has a paramilitary civil protection structure that proved too rigid, too authoritarian, too small in scale and too poorly trained in civil emergency management to tackle the problems of large numbers of survivors (at least 200 000 of whom were made homeless by the earthquake) and rescue many of the live victims who remained beneath the rubble of collapsed buildings. The high death toll also lent weight to the observation (Alexander 1996) that people are more likely to die in earthquakes at night when sleeping, both because vernacular housing is often a major source of vulnerability and because people's ability to react instantly to the tremors is limited by sleep. In fact, the Izmit earthquake struck at 3.02 a.m.

A further approach to the question of vulnerability is shown in Figure 2.3. In this, level of vulnerability is related to level of economic development. The

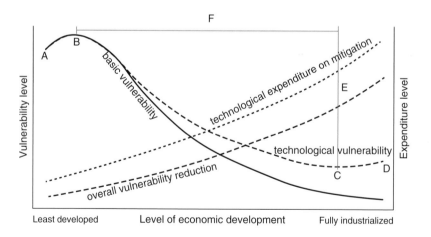

Figure 2.3 Vulnerability in relation to level of economic development and the mitigation gap.

---

19. Much conventional thinking in this sphere is based upon the classic work of Simon (1956); see Whyte (1986: 252).

poorest societies have few resources and few opportunities to reduce vulnerability (point A). As level of economic development increases, then assets at risk
are likely to grow faster than attempts to mitigate the risk: maximum vulnerability will quickly be reached (point B). Thereafter, as economic development
increases, so does mitigation, until minimum vulnerability is reached (point C),
although in highly developed societies further growth puts assets in danger
faster than risk can be ameliorated, and so vulnerability rises again (to point D).
At this point, the vulnerability induced by investment in technology takes over
from that induced by basic living strategies, as the latter has been largely ameliorated. In Figure 2.3 the line from B to E represents the potential for reduction
in vulnerability that can be achieved with existing technology, provided that all
reasonable efforts are made to reduce the risks. This line (F) is the development
gap in mitigation, as it defines what the poorer societies cannot afford in relation
to what the richer ones can.

In synthesis, although chance factors may obscure the relationship between
cause and effect, risk is an unavoidable result of the existence of vulnerability,
and loss is an inevitable consequence of the presence of risk.[20] The presence of
hazard gives rise to both risk and loss, and it is the first determinant of vulnerability.

Vulnerability is also partly the manifestation of the human tendency to defy
hazard – by necessity (lack of alternatives), by default (ignorance) or by wil-
fulness (desire to take risks). It is essentially a deterministic property, whereas
risk can be viewed more in terms of probabilities, although actuaries would tell
us that these are sufficiently estimable to be far from purely random. Paradoxically, this has led to a short-term situation in which it seems, inevitably, as if vulnerability must deepen, threat must transform hazard into disaster, and losses
must become greater (Berz 1988, Kunreuther & Roth 1998). If this is true, the
processes involved in natural hazards may follow a trended metastable function: a path that rises or falls consistently, but with occasional interruptions to
its progress (e.g. Fig. 2.4). Each time disaster strikes, it is followed by a sudden
outburst of legislation and mitigation, which temporarily reduce the risks, but
is soon overtaken by forms of unprotected development that cause the risks to
increase again. Little theoretical work has been done on factors that reduce and
stabilize the upward trend or, in other words, on how to turn the positive feedback of self-reinforcing processes that continually increase vulnerability into
the negative feedback of self-regulating processes that keep it under control.
Metastability and the growth of insecurity are likely to endure while current
trends in population growth, economic development and social expectations
persist. The salient question thus becomes: what scale of human or economic

---

20. See Blaikie et al. (1994) for a more pragmatic and detailed assessment of these links.

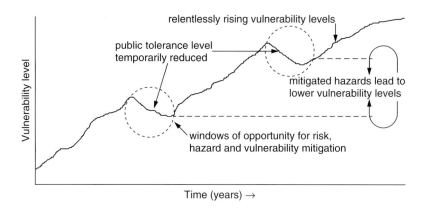

Figure 2.4 Windows of opportunity for the reduction of vulnerability.

losses will be necessary to provoke the kind of reaction that both ends the positive feedback in vulnerability and forces losses to stop rising?

This brings us to the definition of our central term, **disaster**. Its meaning has been intensively debated elsewhere (Quarantelli 1995, 1998) and researchers have gained some novel insights into what the phenomenon signifies. For instance, disasters may be regarded as social catalysts because they can be seen as destroyers of faith in the ability of institutions to protect the public (Horlick-Jones 1995a). The root of the word **disaster** (Latin: *dis | astrum*) can be translated as "ill starred" or "evil star". Synonyms include calamity and catastrophe (the *Oxford English dictionary* definition: "an event that produces a subversion or a sudden and violent change in the order of things"). Crisis and emergency are similar words. Emphasis is thus placed on sudden change, often of an unexpected kind and usually deleterious.

It is clear that the term disaster is multifaceted and open to a range of different interpretations. Its meaning may differ with the use to which it is put. Fundamentally, disasters are abrupt shocks to the socio-economic and environmental system (Burton & Kates 1964; Turner 1976: 755–6). They involve widespread destruction of property and generally substantial numbers of casualties, although no minimum threshold values exist for either problem (Foster 1976). Alternatively, disaster may be viewed as a sudden loss of continuity in socio-economic processes, and as something that slows the pace of development. In economic terms, disasters are a charge upon society, but a burden that is unequally placed, an ill distributed tax on society. As disasters often hurt the poor and the disadvantaged most (Susman et al. 1983, Albala-Bertrand 1993), they offer no opportunity for moral redress: in aggregate, the rich suffer less and recover more quickly from catastrophes than do the poor.[21] For the poor, disaster is often merely one of many afflictions that constitute the harshness of life. For traditional religious people, disaster may be viewed as a punishment sent by the Almighty,[22] an invitation to atonement. In any case, it is often cause for reflection on morals, mores and traits.[23]

Although, after 75 years of social research, prevailing wisdom on disasters is firmly established, it is wise to question it from time to time. In social terms, a disaster is a *non-routine* event but a *routine* social problem,[24] because disasters are recurrent and because they can at least be anticipated, even if they cannot be predicted. That much is not in doubt. It is also clear that disasters expose the ills and weaknesses in society, although they may also emphasize its strengths. With regard to the latter, research has shown that the short-term response is one of increased social cohesion and thus of mutual solidarity (Barton 1970, Sweet 1998). But is this always so? Is it quite correct to expect looting and other forms of anti-social behaviour, or panic and other forms of asocial behaviour, to be as uncommon in disaster as they are made out to be? If, as some researchers have suggested (Hewitt 1995), natural catastrophe is not really generated by geophysical agents (which are mere triggers), but by the underlying weaknesses in society, is it realistic to assume that there will be a substantial positive reaction during the aftermath? The debate remains open: on the negative side, societies that are in a state of serious disequilibrium before disaster strikes are unlikely to rectify themselves when it does, and thus are liable to manifest anti-social and asocial traits; but on the other hand, there is no doubt that society is extraordinarily resilient in the face of death and destruction. Thus, disaster is usually juxtaposed with *resilience*: it is a convulsion in the social system but not necessarily – indeed, not usually – a decisive one (Hewitt 1983, 1995, Ebert 1997).

Not only are disasters hard to define in terms of thresholds, but there is also a taxonomic problem, as there are no sharp boundaries between natural calamities, technological disasters (e.g. transportation accidents and toxic spills), social calamities (e.g. riots and terrorist outrages), and the devastating effects of armed conflict. Indeed, there is much overlap in the details of these phenomena (Quarantelli 1995).[25] Although this renders classification difficult, it does present many opportunities to transfer knowledge of one type of phenomenon to another.

---

21. This was clearly demonstrated in Nicaragua in the wake of the 1972 earthquake that laid waste to parts of Managua (see Ebert 1981).
22. See Sims & Baumann (1972) and Baumann & Sims (1974) for important studies that incorporate this perspective. See also Nurul Alam (1990) and Mamun (1996).
23. See Anderson (1967) and Gherardi (1998) for the cultural side and Beatley (1989) for the moral aspects.
24. But see Drabek (1989).
25. Hewitt (1997) preferred to treat hazards in relation to key concepts, such as endangerment, insecurity and vulnerability, rather than through the classic taxonomy based on the distinctions listed here. There is much to be said for this approach, as it aids free comparison between the effects of very diverse agents of disaster.

In synthesis, disaster is what its victims and participants perceive it to be: it is the sum of many personal catastrophes, but with a certain gestalt, as it also represents a shock to the social organism. To those who are versed in the subject and have lived through a major disaster, study both illuminates and destroys the phenomenon, for academic methods cannot analyze disaster as a whole, only break it down into its constituent parts. To the people who are caught up in it, disaster is, however briefly, an all-embracing phenomenon.

Having considered the question of how disaster and some allied concepts are defined, we will now examine how the definitions have been put to use in the study of extreme natural phenomena and the human response to them. The next chapter will therefore trace the development of certain ways in which human societies have learned to adapt themselves and their environments to natural hazard risk.

CHAPTER 3

The study of disaster

This chapter begins with a critique of alternative models of hazard management. Given the role of academic research in stimulating societal adjustment to hazards, and other forms of adaptation to risk, it will then be appropriate to take a critical look at how researchers have set about studying extreme natural phenomena and their consequences. As some of the shortcomings of academic study have resulted from the difficulty of obtaining data of adequate quality, this problem is considered in some detail. One consequence of the unreliability of data is a lack of studies and models that express the geographical pattern of disasters. We examine why this is so, because it is an important shortcoming of academic work in the field in question. The present chapter closes on a more optimistic note with a re-evaluation of the human ecological approach to catastrophe, after more than three quarters of a century in which it has been employed to reveal how people are affected by environmental extremes.

The evolution of approaches to natural disaster

A classic model of the evolution of human adjustments to natural disaster has emerged (Kates 1971, White 1973, 1974, Burton et al. 1978, 1993). It can briefly be described as follows. Initial settlement of the hazardous area has low density and slow rates of urban growth. Protection against natural hazards is generally lacking or minimal, and disasters cause damage that is gradually repaired or substituted by rebuilding. Neither the cultural nor the technological underpinnings of society favour energetic mitigation, although a form of hazard aversion often emerges in the selection of sites for buildings and towns. As agriculture requires both an intimate knowledge of land capability and slow painstaking development, it is often the only field in which mitigation is practised, through, for example, modest efforts at flood or erosion control. With the passage of time, industrialization or increasing urban-economic growth occur. Untrammelled expansion of human activities increases the exposure to risk of both people and fixed capital. Hence, the toll of disaster impacts rises. Initially, the response to disasters is simply to refinance development by giving unconditional grants and loans, a practice that has been termed "forgiveness money", because it does nothing to curb imprudence (Kunreuther 1974, Sorkin 1983, Burby et al. 1991). Reconstruction therefore takes place with a fair degree of geographical inertia – the failure of disasters to dislodge human settlements from vulnerable sites – which may prolong or even extend vulnerability to natural hazards.[1] But at a certain point a threshold of intolerance is crossed and structural protection is instigated in a major way. Its extent and level of sophistication correspond to needs identified by losses incurred in past disasters. At the same time, further urban and economic growth occurs, and more expensive and extensive structural protection is required. Growth and hazard mitigation thus exist in an uneasy state of mutual sustenance; the former stimulates the latter in an intermittent sort of way, but erroneous assumptions about the efficacy of structural protection are often used as an excuse to permit further growth. Thus, for example, the building of levees on the Mississippi River led to increased urbanization of their floodplains under the assumption that the protection was sufficient (Belt 1975, Myers & White 1993, Changnon 1996). In this and other cases, rarer higher-magnitude impacts lead to greater disasters (as the levees are overtopped by flood levels that are not foreseen).

The first response to the failure of structural protection is usually one of renewed investment in it: higher levees, more dams, deeper relief channels, and so on. But structural protection inevitably has its limitations, the results of which are manifest in higher disaster losses, and so the next recourse is to nonstructural protection. This can encompass a mixture of approaches, including warning and monitoring, evacuation procedures, civil protection and disaster management, norms and codes, land-use control through planning instruments, and insurance and financial incentives to encourage hazard mitigation. Broadly speaking, the measures can be classified variously as incentives, restrictions and first-aid procedures (White 1973, Godschalk 1991). Four types of problems tend to be encountered with such approaches: opposition on the part of libertarians and opportunists who would resist government interference in their activities, failure to observe enacted norms and laws, inability to finance agreed measures, and failure to organize an effective response to the hazards. We might add that there are also risks of encouraging a population to depend on measures that have been determined outside the community, and often to rely on unstable sources of funding, which can lead to aid-dependence or "assistentialism", as

---

1. For a description of geographical inertia, see Alexander (1993: 5).

it is sometimes ironically known.[2] Moreover, it is surprising how rarely cost–benefit approaches are used to define mitigation strategies, even though common sense rules that scarce resources must be invested wisely in order to obtain the maximum mitigation benefit (Ouellette et al. 1988, Britton & Oliver 1997).

The end product is a package of mixed structural and non-structural mitigation measures, in which damage is reduced and lives are saved by a pluralistic approach to coping with the hazard.[3] Currently, there is also a trend towards an "all-hazards" approach, in which protection is extended to counter a range of risks that might significantly affect each site (Perry 1985, Rosenthal & Kouzmin 1997).

In the twentieth century, progress has generally been equated with economic and technological growth. Industrial research and development have fuelled rises in gross national product and gross domestic product, and these have been widely regarded as indicative of human wealth and wellbeing. Although natural hazards can act as a significant brake upon wealth-generating processes – especially in small, poor countries with low and precarious growth rates – in no case have natural disasters put a stop to economic growth. In economic terms, they are mere irritants, perturbations in the flow of capital.[4]

The primacy of the economic growth model in the affairs of nations has in large measure conditioned the approach to hazards. Structural mitigation is preferred for obvious reasons by the construction and economic growth lobbies. Technological hardware production, fruit of the "military–industrial complex", as that agglomeration of industries and politics has come to be known, has offered ever more complex, expensive and sophisticated solutions to the problem of hazards.[5] Yet growth at any price represents a potential source of vulnerability.

Consider, by way of example, the central Italian city of Perugia. This regional capital, and commercial and industrial centre, spreads its 145 000 inhabitants across a series of hills and valleys, some of which have unstable slopes. In the 1950s, residential development spread up slope from the main railway station,

---

2. However, Bradbury (1998: 332–3) questioned the existence of this phenomenon.
3. In an earlier work (Alexander 1995b), I suggested that the usual progression from structural to mixed structural and non-structural approaches to disaster may be inappropriate for some developing countries, which could more efficiently limit their dependence on expensive engineering measures and pass directly to an increased reliance on non-structural approaches.
4. This view is both confirmed and contested by the excellent historical review of the economic impacts of disasters in Europe and Asia given by Jones (1987). A contrary view was expressed by Shah (1995), who suggested that a repeat of the 1923 Tokyo earthquake could jeopardize the world's financial stability.
5. For an analysis of the consequences of this in developing countries, see COPAT (1981). See also Albala-Bertrand (1993) and Pugh (1998).

which is situated on the valley floor beneath the city. No account was taken of the fact that in between the fourth and first centuries BC the Etruscans grappled unsuccessfully with the problem of slope stability, their deep drainage channels, or *cuniculi*, having failed to rid the area of its excess soil moisture. As a result, slow landsliding began to shift houses and apartment blocks on their foundations until a total of 80 buildings were affected. At this point, in the 1970s, it was decided to build a large new commercial and banking centre on the site. A technological fix-up was urgently required and so a system of 240 linked drainage wells was designed and installed (Bianco 1986). Automatic piezometers now monitor soil moisture, and pumps drain the slope, although initially much of the monitoring had to be done manually once a day (Righi et al. 1986). While the system continues to function there is little risk of slope movement, or so one hopes, but huge investments in property have been made on the basis of a technological system for landslide control that is certainly vulnerable. Only hindsight will determine what true level of confidence can be placed in it. In this case, as in so many others, science has acted as handmaiden to an alliance of property developers, architects and politicians. Time will tell whether the development in question is a folly or of net benefit to the city.

Somewhere in the world each year technologically sophisticated loss-reduction mechanisms fail to live up to their estimated potential. Overall, the pattern is one of falling casualty totals but rising economic losses.[6] Technology does indeed help protect lives – at least in the aggregate it does – but can it protect itself? However, in fairness it should be added that there is also a trend towards loss inflation by fiscal mechanisms. To begin with, losses are now better accounted for than at any time in the past; there is a much greater tendency to take into account the hidden aspects, for example to cost and quantify loss of amenity, sales or earnings, even loss of time, not merely loss of property.[7] Secondly, insurance claims are easily inflated, especially where liabilities can be contested vigorously by policy holders.[8] Thirdly, post-disaster expenditures are often aimed at doing far more than simply restore pre-disaster conditions. Nevertheless, there has been a general failure to install a "culture of civil protection" with respect to loss minimization, whatever benefits have been gained in terms of the protection of lives.

Little or no evaluation has been given to the technological approach to mitigation in terms of any alternatives (Alexander 1995b). But there are some startling potential conclusions. For instance, in the 1960s and 1970s, China could

6. See the annual reports of the International Federation of Red Cross and Red Crescent Societies for statistical details (e.g. IFRCRCS 1994, 1996, 1997, 1998).
7. See, for example, Tubbesing & Mileti (1994).
8. See Berz (1991) for a general view of natural disasters in relation to insurance and re-insurance.

not afford, and was not technologically advanced enough to devise, a national system of earthquake monitoring and prediction based on expensive scientific methods. Instead, it opted for a program in which tens of thousands of ordinary people were appointed as observers and trained to recognize earthquake precursors – turbid wells, fireballs, lightning, unusual animal behaviour, and so on (Bennett 1979). Chinese seismologists conducted long periods of continuous fieldwork in areas of high earthquake risk. The Chinese authorities later claimed that the sheer volume and systematic nature of the observations enabled 11 significant earthquakes to be predicted in advance, including the magnitude 7.5 Haicheng event, which caused very serious destruction (Kisslinger 1974, Cheng Yong et al. 1988). During the same period, the Japanese invested US$2 billion in studies of the Earth's magnetism alone, although magnetic fluctuations have not yet proved a reliable earthquake predictor (Rikitake 1984, Scholz 1997). Likewise, US investment in scientific monitoring was high, although tangible results remain elusive to this day.[9]

However, this admonitory tale should not be interpreted as a call to return to alternative technologies and an artisan's approach to the problems of big science. It should instead be seen as a call for a more intuitive approach, in which the solutions adopted are not determined by previous investment strategies and technological lobbyists.

It should be clear, at least, that the standard model of how disaster mitigation strategies evolve is often inefficient. Let us therefore consider the prospects for an alternative and more efficient approach. To begin with, they are probably most favourable where the sequence described above is still at an early stage and approaches to mitigation have not acquired an inertia that will trammel them. The dictates of prudence and the need to foster grass-roots democracy demand that one proceed from non-structural to structural protection, not vice-versa (Wisner et al. 1977, Alexander 1995b).

The traditional model of hazard mitigation proceeds from large-scale overall planning to detailed efforts at particular local scales, where the results tend to be imposed upon local communities by circumstance. One alternative is to let the need for structural protection be determined in the context of natural regions.[10] Within these, local initiatives should be promoted in a uniform aggregative way. Some needs must be met at the overall scale. Examples of these are river basin management and flood control,[11] anti-seismic norms and retrofitting

---

9. However, Silver & Wakita (1996: 77) noted that the 1995 Kobe earthquake was retrospectively found to have been preceded by a variety of clearly identifiable geophysical changes. They were simply not noted until after the disaster, which highlights the importance of social organization (or perhaps merely inspiration), as well as scientific method, in such research.
10. Bioregions might be used (Alexander (Donald) 1990).

(Ambrose & Vergun 1985, Coburn & Spence 1992), and volcanic hazard monitoring (Tilling 1989, McGuire et al. 1994). Other initiatives can be tied to local needs in such a way as to maximize the protection of individual communities.

One problem here is that the concept of "natural regions" is at best hazy and at worst fraught with difficulty (Alexander (Donald) 1990). In essence, five types of geographical region concern us here: bioregions or ecoregions (e.g. the Connecticut River Valley in New England), political regions (e.g. Kuwait), ethnocultural regions (e.g. Cataluña), technoregions (e.g. Silicon Valley) and hazard regions (e.g. the San Andreas fault zone, California, or coastal Dade County, Florida). It is necessary to find the best possible fit, or compromise, between the five entities in order to define the operative scale for basic hazard mitigation. The task is not easy.

Commonly, hazard mitigation follows a path from spontaneity to organization, with the latter often being after the event and therefore a fudge concocted to remedy inefficiencies. The alternative requires forecasting of hazards, foresight in preparing for them, and an adequate consensus about the need for prior planning. Increased organization is obtained at the price of restriction on personal liberties. Such a situation is not easily attained, as the power structure of society, with its hierarchies and oligarchies, does not often favour grass-roots groups against the elite (Wisner et al. 1977), and the former seldom control the means of production. It follows that democracy, and not any subversion of it, is at the heart of hazard mitigation.[12] Imposed solutions are implicitly flawed.

Prospects for a new approach to disaster mitigation are therefore greatest at modest geographical scales in areas in which hazard impacts, urban-industrial growth and existing structural protection are not greatly out of step with each other. It is a valid general principle not only that disaster protection must be integrated with general socio-economic growth but that both must be sustainable (Varley 1993). Several problems arise with this. First, it is difficult to define what is sustainable and how to sustain growth. Not only are there varying degrees of sustainability, but as circumstances change so do criteria. Conflicts occur when sustainability must be defined in terms of competing economic sectors, different goals over time (especially in the short and long term), varying spatial scales, and objectives that differ with level in society's hierarchy (Bender 1992). Nevertheless, it follows that mitigation must be carried out in accordance with ecological principles if it is not to create new or enhanced sources of vulnerability (Faupal 1985, Bates & Pelanda 1994). Non-renewable resources can still be used, but the results must be durable and reliable.

---

11. Costa (1978) presented this more as a quandary than as a strategy.
12. See Wolensky & Miller (1983) for a discussion of this point and Blocker et al. (1991) for a field example.

Likewise, disaster mitigation must not contribute significantly to the deval-uation and expropriation of common property resources, the so-called "tragedy of the commons" (Hardin 1968). It can instead make excellent use of public goods (e.g. meteorological forecasting) in which usage does not subtract from availability. Hazards are in effect common drawbacks which it is in everyone's interest to mitigate. The present dilemma for many countries is the extent of public versus private involvement in such efforts. The almost universal ten-dency to reduce the size of the public sector of national, regional and local econ-omies may in the end be a prime source of hazard vulnerability if there are insufficient incentives for the private sector to take up the slack. It is also ques-tionable whether private initiatives contribute enough to the public good, again a question of equity (Abernethy & Weiner 1995).[13]

However the public/private debate is eventually resolved, one cannot deny the role of political stability in promoting hazard mitigation, and the impor-tance of sufficient integration between the various levels and factions in society. In the aftermath of the December 1988 Armenian earthquake, Azeri relief work-ers were stoned by Armenian nationalists (Alexander 1995c). The relief effort rapidly deteriorated into an ideological battle that soon became a bid for inde-pendence which led to a brutal war over territory. As a result, recovery from the earthquake all but ceased (Verluise 1995).

In other cases the key to the problem of mitigation lies in reversing the wide-spread tendency for income gaps to widen between the rich and the poor. This is once again intimately linked to the problems of poverty and marginalization, which so much determine the level of vulnerability to natural hazards.[14]

In synthesis, let us state six principles of good disaster mitigation and pro-tection:

- It must resolve, rather than increase, sustainability conflicts.
- It must not facilitate or lead to development that is unsustainable in terms of the misuse and lack of preservation of resources in future perspective, the eventual failure to achieve morally reasonable objectives, or the ten-dency for resources to be inequitably distributed.
- As much as possible, it must use planning measures to reconcile eco- or bioregions, political regions, ethnocultural regions, technoregions and hazard regions.

13. In 1998 the US Federal Emergency Management Agency instituted "Project Impact", a plan to stimulate local-scale risk mitigation by systematically encouraging cooperation between the public and private sectors. In its first year more than 50 small to medium-size towns and cities adopted the plan and it was judged a great success.
14. See Blaikie (1985) for an excellent discussion of this problem with many regional examples. However, Cannon (1994: 27) argued that there is no absolute correspondence between vulnerability and poverty: for instance, differences in gender and ethnicity may give rise to vulnerable people in less vulnerable groups.

- It must not subvert common property resources to sectoral or private gain at the expense of public security and wellbeing (i.e. it must not lead to the "tragedy of the commons").
- As much as possible, it must enhance the value of public goods.
- It must be part and parcel of development that is appropriate in scale and degree of sustainability to its environment.

Academic studies of hazards and disasters

As academic studies have had a significant impact in determining hazard and risk mitigation policies (White 1973, Yin & Moore 1985), it is opportune to consider the organization, history and present condition of disaster studies.

In the study of disasters and mass emergencies there are at least six schools of thought (Alexander 1991a; 1993: 12–14). These stem variously from geography, anthropology, sociology, development studies, medicine and epidemiology, and the scientific and technical disciplines such as volcanology, seismology and engineering. In a few cases there are some strong links between the schools, for example, between anthropology and development studies; and naturally there are strong links between the various branches of the technological disciplines, and between those disciplines that deal with monitoring hazards, such as remote sensing and seismology, and those that deal with the built environment, namely engineering and architecture. But it is striking how the various disciplines view the same phenomena in relative isolation from one another. Although there have been some noble attempts to bridge the yawning gaps between the vantage points (e.g. by studying the psychology of how buildings are used during earthquakes),[15] specialists from various disciplines appear to have made little effort to appreciate alternative points of view. Perhaps they cannot make the necessary transition in thinking. Or perhaps the appropriate training in mental reorientation is lacking. Alternatively, they may fear loss of credibility by stepping into a new field (and, of course, no *bona fide* scientist wants to be regarded as a charlatan). At any rate, disciplinary barriers have impeded progress towards a better understanding of emergencies and how to manage them.

In order to understand this parcelling-out of the field, it helps to classify the compartmentalizers. Just as history cannot fully be understood without some understanding of how it is perceived, how the great welter of historical facts is

---

15. See Taylor (1984), or, for the relationship between geomorphology and architecture with respect to landslide damage, Alexander (1989).

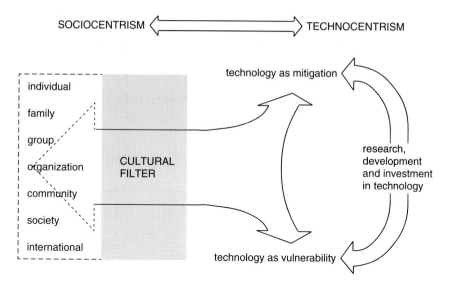

Figure 3.1 The cultural filter through which technology is interpreted in response to natural hazards.

shaped into a coherent account of events (or in other words, what historiography is), so the activities of other academic groups need to be viewed in the light of the sociology of knowledge – the organization of the organizers of knowledge. One is tempted to start by dividing the students of catastrophe into "technocrats" and "humanists", following the much-used distinction between those who would manage the environment by applying doses of technology and those who prefer to let nature have its way as much as possible. But in reality, attitudes to disaster tend to be more complex than a straight division between technocracy and ecology. Some indications of why this is are given in Figure 3.1. Hence, in the social and humanistic disciplines that have occupied themselves with disasters, no such distinction has arisen. Indeed, although one might expect the social scientists to be "greener" and less wedded to the so-called "technofix" approach to disasters, it is not so: the bulk of the literature in sociology, psychology and geography, and a fair proportion in anthropology, stems from a very conservative epistemological tradition in which problems are regarded as best solved by human ingenuity, which provides evidence of the primacy of our species over our own habitat (Pepper 1984, 1996).[16] This gives rise to a curious paradox, as the very same literature has consistently proved that human ingenuity is itself the culprit in disasters: the overdeveloped coasts with

---

16. It has also been suggested that traditional sociology, born with the Industrial Revolution, was developed to justify capitalism, not to criticize it (Anderson 1997: 4).

their hurricanes and tsunamis, the aseismic houses, and the excessive demand for water in areas of periodic shortage – to name but three examples from among many. In truth, a radical critique has emerged, first in France and then in North America and the British Isles. In this, the direction of causality is reversed, so that human actions, rather than geophysical extremes, are considered to be the real culprits when the ground shakes, the volcano erupts, or the river overflows (Hewitt 1983). But this aetiology has not become the accepted norm.

Perhaps there are two root causes of the present fragmentary nature and slow progress in disaster studies. The first is fear of loss of identity and the second is a question of funds and power. The two are strongly linked, in that funding confers visibility and hence identity upon researchers. Conspicuous and consistent success in funding research projects opens the door to better working conditions, more respect, and a better job in a more prestigious institution. A strong sense of identity is the first necessity when marketing a research proposal. Interdisciplinary research is generally fuzzier in terms of its aims, progress and outcome than conventional discipline-based investigation. Moreover, research proposals may well be judged by a panel of experts, who are almost inevitably conservative in outlook when it comes to giving away money and who will view the proposed research in terms of the competitive divisional sectarianism that characterizes so much of academic and professional life. Sadly, in the current triumphalist, monetarist, consumerist epoch, the most eloquent advocate is money. Throughout the world, funds are now most easily attracted to research where the outcome is an applied fix-up with tangible results, moved earth, strengthened concrete, or stainless steel instrumentation – the technofix solution again.

Over the past century, science has mobilized itself to serve the process of economic development, smoothing the path, solving the problems, facilitating what has come to be known as "progress". Hazard and risk mitigation are part of this endeavour, but they tend to be secondary components. Thus, engineering gives us taller buildings, longer bridge spans and larger loads, but only when collapses occur on a spectacular scale does seismic engineering give us a building, a bridge span, or a load-bearing support that will resist the horizontal acceleration caused by a major earthquake (Zebrowsky 1997: ch. 3). The invention of reinforced concrete, machinery, the internal combustion engine and prefabrication have facilitated the rapid expansion of urban growth into areas of severe natural hazard. Rare indeed are the cases where the same inventions have been used artificially to create enough safety to justify such expansion in terms of risks that have been sufficiently abated. In this particular respect it is striking how little research and investment have gone into *limiting* human aspirations to cohabit with nature's extremes (Burton & Hewitt 1974, Faupal 1985, Bunin 1989, Bates & Pelanda 1994). Are we too cocksure a race, or are we simply unable to perceive the benefits of wanting less?

Science, principally *hard* science, and engineering have worked the technofix approach to hazards into the position in which it rests: not merely a dominant, but practically an unchallenged, orthodoxy. Mitigation of hazards therefore *means* abating them wherever possible or building structural defences in places where nature's forces cannot be lessened. Only, when this has been done can other approaches be considered. Thus, it is hardly surprising that there is generally a progression from no mitigation to only structural measures and to mixed structural and non-structural approaches.

Strangely, the reaction to this among those disciplines that deal with social issues has been muted and uncoordinated. They have dug their own burrows out of the way of the funding melée. One suspects that the principal reason for this extraordinary failure of nerve lies in the same fear of losing one's professional identity that keeps the technological fix-up solutions so closely aligned with their respective disciplines of origin. This fear is especially strong in sociology, psychology and anthropology. Practitioners of these disciplines continue to insist that there is enough, or indeed *too much*, technology in the field and the real solution to humanity's disaster problems lies in understanding and coping with the social issues. Quite so, but it is striking how weak the combined voice of the social sciences is (except, as in the case of geography, where it has lent support to the technocrats). In point of fact, territoriality has muted it. Sociologists dare not admit to the anthropological perspective, anthropologists scorn the sociological approach and, when psychologists describe the same phenomena as do social researchers, it is as if the two groups were discussing entirely different things.[17]

The gap seems to yawn widest between sociological and anthropological approaches to community emergencies. For fear of losing their identity, or so I would argue, the sociologists dare not acknowledge the weight of culture in shaping social issues, whereas social anthropologists dare not admit that much of the problem lies purely in social and organizational relations between individuals. Thus, we end up with not one but several beleaguered minorities in the quest for prestige and authority in the study of human extremes. It is sad to watch, for the subtleties of human relations are less colourful and less pictorial than a spectacular building collapse or lava fountain – the direct preserves of the world's technocratic scientists.

Another reason for the siege mentality that has developed in social studies of disaster lies in the predominance of Anglo-Saxon culture (and particularly North American attitudes) in the social science literature, although perhaps a little less so in anthropology than in sociology.[18] The dominant cultural matrix tends to

---

17. See the debate between Torry (1979) and Burton et al. (1981) for an entertaining academic fireworks display sponsored by practitioners of rival disciplines.

be mechanistic and individualistic, and sometimes crudely so. It has little time for abstract views of the human condition and thus often tends to despise and denigrate spirituality, for example, as mere fatalism. It minimizes the weight of history in shaping attitudes and actions in the face of disaster. Attempts to broaden the approach have often amounted to little more than extending it into societies that share enough of the usual attributes not to invalidate the established precepts or any significant conclusions that arise from them. Hence, although Japanese society – much studied in terms of how it copes with disaster – is undoubtedly much less individualistic than its North American counterpart, it is nevertheless mechanistic in much the same way: it thrives on the technofix.[19]

In all this, the silence on social issues in Third World disasters is sometimes deafening. The neophyte reader might be led to assume that there is something rudimentary and simplistic about social relations and reactions to disaster in developing countries (as if all nations thus classified were alike!). Poverty, low GNP and lack of resources might be treated as problems that make social and cultural relations primitive and rudimentary, and action and reaction to disaster simple, and thus not worthy of much study. A few dedicated social anthropologists have worked hard in the Horn of Africa (Turton 1991) and the Andes (Oliver-Smith 1977a, 1986) to dispel this myth, but their work has had little impact on the world view of disaster researchers in the world's richest countries.[20] Society and culture in the USA are highly self-absorbed. Other countries look to the USA for academic leadership and usually follow its initiatives (Nemec et al. 1993). As a result, generalizations are made about human behaviour under duress as if they applied to the whole of humanity, *but no one knows whether they do*, a convenience born of ignorance is in danger of becoming accepted orthodoxy, a truism.

The only way to find out the truth is to view the societal impact of disasters in terms of two fundamental determinants: history and culture. To take an extreme example, when disaster strikes Santa Barbara, California, it impacts a community whose collective existence largely dates from no earlier than 1890, whatever the histories of its individual residents might be. Yet when catastrophe occurs in Andhra Pradesh, Sicily or Sichuan, it impresses itself upon a social fabric that has taken many centuries to mature. Older societies carry with them the half-remembered traces of conflicts, alliances, social groupings and struggles. Events are transformed into history, history is absorbed into culture, and that provides the matrix for reactions to disaster. When disaster occurs, it

---

18. See, for example, the excellent work of Anthony Oliver-Smith for some detailed studies of non-Anglo-Saxon disaster cultures (e.g. Oliver-Smith 1986).
19. See, for example, Normile (1994). Even some of the social studies (e.g. Yoshii 1989, Palm 1998) appear to reveal a somewhat mechanistic attitude towards human relations.
20. New work in Bangladesh is beginning to change this; for example, see Paul (1999).

contributes to the tapestry of events that make up a people's history. Emigration to the Americas has traditionally been one of the ways of starting afresh by seeking to escape the long shadow, and often the crushing weight, of history.[21] But there are distinct dangers in a social science view of disaster that evokes a collapsed view of history. To begin with, it allows little opportunity to generate the methodology needed to isolate and study the historical influence on our attitudes to danger, emergencies and disaster impacts. Most fundamentally, it offers a model in which essentially ahistorical peoples and societies become the norm, and any deeply historical culture is reduced to the status of an appendage. The result is a mechanistic view of society, one that is fertile soil for the implantation of technological solutions to disaster, but barren ground for the resolution of age-old conflicts and ancient rancours.

The failure of nerve in the social sciences has had two principal effects. One has been the lack of ability to generate a body of theory that adequately unifies society with culture and history, and the other has been a neglect of the study of the impact of science and technology on society. The latter is an essential counterbalance to the technocentric view of problem solving. In many instances increased investment in expensive technology for monitoring and mitigating hazards has been followed by increasing costs and losses when disaster strikes. This has been the case with, for example, the construction of flood defences on the Mississippi River (Changnon 1996) and on the Ganges and Brahmaputra rivers in Bangladesh. In the former case, the tireless efforts over half a century of the geographer Gilbert Fowler White have succeeded in creating a new role and status for non-structural flood mitigation (Whyte 1986) but have not abated the demand for structural protection. In the latter instance, ridiculous engineering projects have been proposed but gradually modified by the weight of scientific opinion against them (Boyce 1990, Paul 1999). A body of knowledge on the social science aspects of flood (and drought) hazards in Bangladesh has begun to form, using White's school of thought as its initial model. But it is as yet unclear what theoretical form the final portrait of society, its aspirations and fears, will take, and whether this will resemble the North American model (Alexander 1995b). That is even more so for the prescribed remedies to the flood problem, for there are distinct dangers in the wholesale transfer of solutions from one continent to another without tailoring them to local cultural conditions.

We are left, then, with a nascent field (a field without a name, as the term "disasterology" rightly has no following) distinguished mainly by its fragmentary nature. From the point of view of the physical sciences, so much is now

---

21. Placanica (1985) noted that after the six earthquakes that devastated the southern Italian region of Calabria in 1783, the newly formed USA set up emigration offices in the villages affected by the tremors. In some places, the departure for the New World of many able-bodied men caused the reconstruction process to stagnate for want of manpower.

known about the spatial distribution and temporal pattern of hazards, not to mention the processes and mechanisms at work, that there are virtually no grounds for pleading ignorance (El-Sabh & Murty 1988, Zebrowski 1997). Yet, accurate monitoring and perspicacious forecasting can do no more than furnish signposts for social actions, economic investment and organizational response. It is easily forgotten that the actual decisions are made on the basis of priorities and preoccupations that may have relatively little to do with objective analysis.

There are two solutions to this problem. First, disasters must be viewed as holistic problems, which, rather than being interdisciplinary or multidisciplinary, are *independent* of academic disciplines. Secondly, on the basis of a greater degree of unification among the social sciences and humanistic disciplines, a new and more circumscribed role must be defined for the physical and monitoring sciences. In this it is important to recognize that most people do not have a clear idea of what science is and what it is capable of achieving (Lindell 1994, Larson & Enander 1997). Their aspirations – at least on a superficial level – are easily diverted by technology, but they are not moulded by the aims and objectives of scientific rationalism. Hitherto, the scientific solution to disaster and its problems has simply been imposed on the populations at risk. Social and cultural studies are capable of mediating such a situation in the interests of democracy, but only if there is a greater rapprochement between the social and the physical sides of academic study.[22] This requires a radical reorientation of the predilections and techniques used in training social and physical scientists and in validating their competence.

On the unreliability of disaster data

The practical and academic study of disasters is inhibited by the incompleteness and unreliability of basic statistical data. We live in an age in which great emphasis is put on numerical information. People hunger for it, it confers respectability on news reports, it reassures people who are uncertain, convinces the sceptical, and carries the argument, yet we do not extend quite the same level of interest to the question of how accurate it is. But if we are truly to understand natural disasters and to appreciate the trends behind them, we need basic information that is precise and reliable.

To keep the matter in perspective, it must be said that various institutions do endeavour to collect comprehensive information on contemporary disasters.[23] Their efforts are aided by the Internet, which has vastly improved accessibility

---

22. Nemec et al. (1993) took some tentative steps in this direction, but much remains to be done.

to news and basic data on natural catastrophe (Spurgeon 1996).[24] Sources are now more numerous, access is more rapid and there is also the possibility of automatic delivery of bulletins and situation reports using Listserve utility programs (i.e. programs that automatically send messages from contributors to other subscribers). Nevertheless, assembling, collating and checking data are still hugely time-consuming processes that inevitably force the archivist to form criteria in order to define what information should be preserved and in what form and format. Current databanks thus tend to restrict coverage to the aspects that most interest their creators; for example, the Red Cross is mostly interested in data on casualties and human suffering, whereas the re-insurance industry is mainly concerned with the ratio of total monetary losses to insured losses. Thus, there is no comprehensive, freely available databank and, moreover, no system of controls is ever applied in order to assess the accuracy and reliability of compiled data.[25]

In many fields, from economics to public health, from raw materials to demography, it has for decades been routine to collect data on a country-by-country basis, according to world standards, and to publish them internationally. It is therefore surprising not only that this is not done for disasters but that there are no generally agreed international standards or procedures for reporting such data. Yet it need not be so. When disaster strikes, governments are almost invariably the first to assess the situation and compile lists of casualties and damage, even if they do so with varying degrees of precision and application. It is but a short step to formalize this procedure and create a supranational body that would bank the data, ensure their accuracy and make them generally available.

The first question, however, is what data? Categories should include death and injury tolls, numbers of people made homeless by the disaster, and levels and costs of destruction and damage (Choudhury & Jones 1996). But although it is easy to list the categories, it is very difficult to define them rigorously, especially in a manner that will suit all uses and circumstances. For example, no category would seem more absolute than death, yet it is not so clear. If death occurs as a direct and immediate consequence of the disaster, there is no particular problem. But then there are indirect causes, such as disease, accident or secondary disaster (Alexander 1985; 1993: ch. 7). There is also a risk of not excluding background mortality, meaning that which would have occurred anyway in the

---

23. These include the International Federation of Red Cross and Red Crescent Societies (IFRCRCS 1998), the Munich Reinsurance Company (Berz 1992), the Disaster Study Unit at the Catholic University of Louvain (Sapir & Misson 1992, Sapir 1993), the Disaster Research Centre at the University of Delaware (Quarantelli 1997), and the Disaster Prevention and Mitigation Unit at the University of Bradford (*Disaster Prevention and Management* 1991–).
24. However, it has not always improved the quality and reliability of data.
25. However, see the discussion of this problem in IFRCRCS (1994: 142–3).

THE STUDY OF DISASTER

absence of disaster. Finally, there is the vexing problem of timing: not all disaster-related deaths occur immediately disaster strikes (e.g. people may die of their injuries at a later stage) and the risk of mortality may remain high if the hazard does not abate. Yet the longer the period over which the dead are counted, the greater the probability of including extraneous cases in the tally.

Nevertheless, it is easier to arrive at a robust estimate of deaths than it is of injuries. Here the problem is complicated by matters of type, severity and cause. Moreover, not all injuries are declared or registered, especially the more trivial ones (Jones 1990). Hence, many injury totals reported after disaster refer only to victims who are hospitalized or who seek treatment in hospitals as out-patients. The solution to the problem of the comparability of injury data lies first in developing a standard set of definitions for types of injuries and the mini-mum severity at which they become accountable. Plenty of research has been carried out on the epidemiology of natural disasters, to the extent that the basic tenets are well known, and measures of injury severity have been developed with varied degrees of complexity and sophistication.[26] But no one has yet developed an international standard injury classification that is specific to any of the natural disasters.[27]

The problem with measuring homelessness is that it tends to be a rather dynamic phenomenon, to the extent that figures can fluctuate by an order of magnitude in a matter of weeks as homes are condemned or demolished, or shelter is organized or survivors are relocated (Alexander 1984, 1986a). How-ever, at any single point in time it is usually assessable.

Finally, destruction and damage are not always clearly distinguishable from one another, as the distinction depends on opportunities for repair and reha-bilitation. Nevertheless, they are generally estimated after disaster, especially if a relief appeal is to be launched, and the resulting data are often of better qual-ity than their epidemiological counterparts. Destruction and damage can be cat-egorized both by severity and according to what is affected: housing stock, pub-lic buildings, industrial premises, artisans' workshops, commercial premises, agricultural enterprises, infrastructure, and so on. Once again, no standard cat-egories have been defined.[28]

In the publication of such data after disaster, there are often predictable patterns of over- and underestimation. These result from a variety of causes

---

26. Compare Baker et al. (1974) with Olser (1993).
27. This begs the question of whether such a classification is truly warranted. I would argue that it is, for the pattern of injuries in floods, earthquakes, volcanic eruptions, hurricanes, tornadoes and other types of natural disaster is broadly predictable (see summary in Alexander 1993: 464–77; also see Noji 1997).
28. But see Hughes (1981, 1982), who has proposed classifications of damage in earthquakes and floods.

including mis-estimation, exaggeration, dramatization and official caution.[29] However, stable estimates are usually made in the end, usually weeks after the disaster impact, although they need to be cross-checked against alternative sources. Of course, by this time, public and international interest in the disaster has probably waned, if not dwindled to nothing, and there may be little incentive to publicize the final set of figures widely. One wonders, therefore, how many of the best data on disasters simply perish unheeded. As no supranational body has general responsibility for the collection and dissemination of disaster data, the data that are collected are incomplete, of questionable accuracy and not necessary freely accessible to all potential users.

Yet the situation could be very different, and it is not an impossible task to turn the tables. One example of what can be done is offered by the work of the World Health Organization in collecting data on internationally notifiable communicable diseases.[30] The benefits of extending notifiability to injuries caused by natural disaster would include a clearer and more accurate picture of the geographical and temporal pattern of mortality, morbidity and loss in natural disaster. With sufficient resources, the same data could of course be collected for technological disasters. Better analysis would be possible and more insight would be generated if such data were available. A convention would be needed in order to define the categories and procedures for reporting data. Thereafter, archival research could probably extend the databank quite far back in time without excessive loss of comparability.[31]

The times are gone when governments regarded natural disaster as a national disgrace to be hidden from outsiders. On the contrary, there are now concerted efforts to improve the level of international coordination in order to fight catastrophe, as in the current plan for an international disaster network under the auspices of G7 countries.[32] It is all the more remarkable that such efforts should be taking place without the firm foundation of a reliable comprehensive database. Perhaps we think we already know too much about disasters?

The unreliability of data is one possible cause of a failure to exploit the potential of the spatial dimension as a medium for modelling. However, there are many other reasons, as the following section will show.

---

29. See Alexander (1996a) for examples of estimation problems with respect to earthquake casualty data.
30. See World Health Organization (1986) for an example of how data collection on communicable diseases relates to natural disasters.
31. Guidoboni (1986) provided a rigorous example of how this can be done.
32. See "Global Emergency Management: an Executive Summary" at the Emergency Preparedness Information Exchange (EPIX) Internet site, http://hoshi.cic.sfu.ca/.

Why there are so few spatial models of disaster

The US government has spent over US$5 million on the development of "Hazards in the United States" (HAZUS), a comprehensive geographic information system (GIS) designed to be used at the local, regional and state level for estimating casualties and losses in earthquakes, and potentially in other forms of disaster (Kircher & Stojanovski 1996, NIBS 1997). It is a sophisticated and flexible system, which can depict many kinds of hazard impact and estimate a wide variety of consequences. Trials in Portland, Oregon, and Boston, Massachusetts, have shown that it adds significantly to emergency management capabilities, especially as it indicates precisely where particular forms and quantities of aid will be required after disaster strikes.

A lively debate accompanied the formulation of HAZUS, particularly regarding the methodology used to estimate damage (Olson & Alexander 1996). For all its virtues this is strongly inductive, which Karl Popper once termed "a slender reed". In order to make it easy to implement, HAZUS supplies some 600 megabytes of spatial socio-economic, demographic, engineering, architectural and geophysical data to the user (who would mainly be expected to be a local or state-level hazard manager or other functionary charged to reduce risks to the community). A more sophisticated and reliable result can be achieved by adding local survey data to the second and third levels of the model, which are designed to be implemented by the user. This is intended to be the task of local technician, as it requires prodigious numbers of field data, meticulously collected, block-by-block (Eidinger 1996a, NIBS 1997). Initial users of the model noted that all this effort to collect and systematize data could lead to results that were spuriously precise (such as predicting 118 deaths in a magnitude 8.0 earthquake, rather than specifying a range of 60–180). It could also lead to excessive concentration on variables that are of limited practical value; for example, in a mass-casualty incident, the number of available hospital beds is not deemed to be as valuable a measure as the quality of care that a patient is likely to receive (Durkin 1995, Levi 1997).

These are typical concerns about methods that are heavily inductive. The alternative – so much more challenging and difficult – is to build a deductive model in which the links and processes in disaster are not passively observed, but are actively specified *a priori*. In reality the process is not so simple, as deduction relies on a certain amount of induction in order to form its premises, and the process of identifying regularities by inductive reasoning often leads to unexpected insights of a more deductive nature, hence the Popper and Hempel loops that formalize ways of integrating the two tendencies. But the points of departure of induction and deduction are radically different, if not diametrically opposed (Harvey 1969: ch. 4).

Induction is the handmaiden of pragmatism. It enables one to fix things without asking too many awkward questions of the data. It also sometimes encourages one to ask the wrong questions, or to simplify them to levels at which they bring forth only the most modest answers. Yet, we are a very long way from being able to conceptualize disasters deductively in terms of a general model based on all four of their fundamental dimensions: time, space, magnitude and intensity (Alexander 1995a). Unfortunately, the allure of digital information processing seems to have led deductive modelling into decline (interestingly, along with uniformitarianism and stochastic forecasting, although the connection is somewhat tenuous). There is little evidence of any attempt to model spatial processes in disaster using deductive principles.

In order to make some very modest progress in this field, it is necessary to start from the most basic level. We begin by dividing the problem into one of modelling impacts, responses and the cognitive world of disaster perception, our own noosphere.[33] In the simplest case, an impact can be visualized as a single, nucleated function that appears on an isotropic plane and represents the direct connection between physical event and human loss. With parameters set by the scope of the impact, the dynamics of diffusion will work in time from the moment of the disaster and in geographical space from the point of impact in radial formation (Fig. 3.2). Distance-decay functions (e.g. exponential, Gaussian and quadratic) can be used to approximate differing degrees of concentration of the impact (Fig. 3.3).

It is obvious that no real territory or spatial process is radially isotropic in this manner (see Fig. 3.4 for an example of one that approaches isotropy), but the model has some validity in simple forms of relative distance estimation. It can be related for example to A. F. C. Wallace's spatial model of disaster, which consists of concentric rings made up of the total and marginal impact zones, an initial zone of filtration of aid from outside, and zones of regional, national and international aid (Wallace 1956, De Ville De Goyet & Lechat 1976). In one study of a mudflow disaster, I found the distance relationships to be roughly logarithmic (Alexander 1986b, 1989c). Wallace's model, which is reminiscent of Von Thünen's characterization of nineteenth-century rural land use, can also be adapted to casualty patterns in earthquakes. Close to the epicentre, deaths predominate; at greater distances they fall equal to and then below the number of injuries; next, only injuries are encountered; and finally only damage without significant casualties. Similarly, the model by Burton, Kates and White of the distributed effects of earthquake (Burton et al. 1978) can be given a three-dimensional representation in which the most concentrated effects are those sustained

---

33. The noosphere is the realm of human minds interconnected and interacting through communication (Svoboda & Nabert 1999).

41

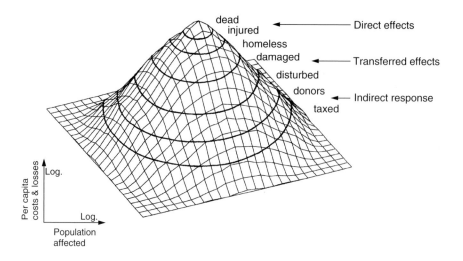

Figure 3.2 Radial isotropic distribution of effects of disaster (after Burton et al. 1978).

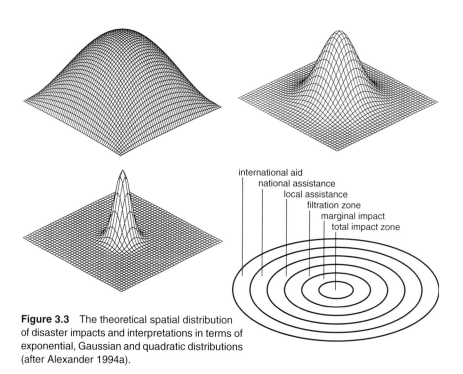

Figure 3.3 The theoretical spatial distribution of disaster impacts and interpretations in terms of exponential, Gaussian and quadratic distributions (after Alexander 1994a).

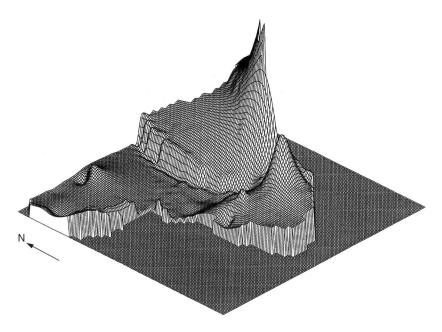

Figure 3.4 Example of radial, sub-isotropic patterns constrained by national boundaries: the distribution of refugees in Africa in the summer of 1992.

by people who are killed in the event and the most widely distributed impacts are upon the general public who pay for the disaster through taxation (Fig. 3.2).

There are two main systematic ways in which a single nucleated impact can be complicated by non-isotropic hazard and non-isotropic response. Both can be subdivided broadly into linear and nodal variations (Fig. 3.5). Among the hazards, linear or axial factors include fault lines, flooded trunk rivers, mudflows, tornado tracks and coastal cliffs. Among the responses are linear urbanization and the role of major routeways. When hazards are considered in terms of nodal or clustered variations, the examples include sinkhole collapses, diffuse centres of volcanic eruption, and episodes of widespread contemporaneous landsliding. The principal nodal elements of response are associated with variations in population density and emergency management capability. As a parenthesis, it is important to note that, in disasters, census data on population density may be a relatively poor guide to actual densities, as aggregate patterns of human activity have to be taken into account and these vary diurnally, from day to day, and seasonally.[34]

The third complicating factor consists of variations that are actually or

34. See Emmi & Horton (1993, 1995) for an excellent example of how this should be taken into account in spatial modelling of natural-hazard risk levels.

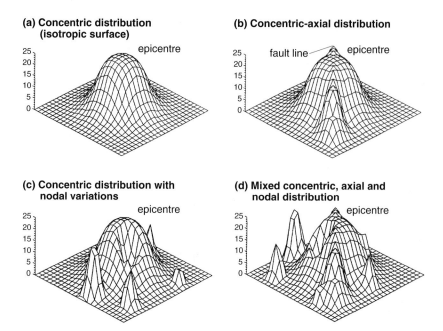

(a) Concentric distribution (isotropic surface)

(b) Concentric-axial distribution

(c) Concentric distribution with nodal variations

(d) Mixed concentric, axial and nodal distribution

Figure 3.5 Nodal and axial deviations from a simple Gaussian distance-decay function in disaster intensity: the example of mortality in masonry constructions during an earthquake of modified Mercalli intensity X (vertical scale: percentage of occupants killed; after Alexander 1994a).

apparently random. These are the sum of each individual aspect of the physical impact and human reaction. For example, in earthquakes the partial collapse of building façades and the incidence of traffic and machinery accidents are usually distributed in a complex manner that is difficult to predict. The same is true of cases of heart attack (Katsouyani et al. 1996), as these depend on a combination of stressful environmental factors (the earthquake and apparent danger that it creates) and predisposed subjects (people whose hearts cannot resist the shock of high instantaneous injections of natural adrenalin). A major earthquake may cause tens of heart-attack victims among the hundreds of thousands of victims of structural collapse, but there is unlikely to be more than a cursory sense of clustering around the epicentre or zone of major damage, as building collapse is not the principal determinant.[35]

In synthesis, for a single, nucleated impact the distributed effects and reactions consist of a smooth distance-decay function on which are superimposed axial, nodal and random variations. If this model is to be used predictively, then geophysical processes must be known well enough to forecast the approximate

35. In this, *perception* of danger is probably more important than *actual* danger (Trichopoulos et al. 1993, Leor & Kloner 1996).

location of the "epicentre" and the instantaneous radial distribution of energy. Geographical factors must be well enough known to plot the spatial pattern of variations in vulnerability, and socio-economic processes must be understood in such a way as to relate aggregate patterns of human activity to risk and actual loss. Specified in this way, the process is reminiscent of one of science's most severe challenges, "white box" systems modelling, in which input, output and process–response functions must all be heavily specified (Chorley & Kennedy 1971: ch. 1). But in reality one can achieve useful results with much less effort and many fewer data. Hence, for example, the regular decay of earthquake intensity with distance from epicentre (Anderson 1978, Brazee 1979, Stillitani et al. 1995): the isoseismals are concentric generalizations of detailed damage functions and they enable one to predict that average damage levels will not surpass certain levels at given distances from the source of the "shaking". Little work has been done to relate this to mortality and injuries in structural collapse, but a relationship certainly exists. A semi-isotropic analogy is that of hurricane windspeeds (which are non-linearly correlated with wind damage) around a landfall that is approximately at right angles to a fully urbanized coastline (Fig. 3.6).

A second area in which deductive spatial modelling would be useful is that of response to disaster. When catastrophe strikes in a relatively limited and well defined area, a series of characteristic movements of people and goods takes place. To begin with, most disasters initiate a convergence reaction (Fritz 1957),[36] in which emergency services, volunteers, relatives of victims, journalists, onlookers and others move into the area in large numbers. Victims who have been rescued alive may be moved out of the area to distant hospitals, and homeless survivors may be evacuated, although in both cases the numbers of those who leave are likely to be small and the probability of eventual return migration will probably be high. Paradoxically, although each of these processes is well known, little general theoretical information exists on the locational and distance relationships involved. However, we may hypothesize two forms of geographical linkage. In the first case, the bulk of aid and relief (perhaps more than 90 per cent) will probably come from the region and nation in which the disaster has occurred. Hence, there will be particularly strong links to regional and national centres of government, which are also likely to be the largest cities. In the second instance, there are likely to be particular links with places that have some special involvement with the disaster area. Emigrants may return to the stricken community from their places of work abroad, former colonial powers may make special donations, and countries that have built up bilateral relationships based on trade or development aid may be among the foremost donors (Borton 1993, Hendrickson 1998).

36. For more recent views, see Scanlon (1992) and Wenger & James (1994).

The spatial relationships thus generated are likely to be more complex than meets the eye. Bilateral relationships can span half the globe, and modern means of transportation and telecommunications can render them almost as effective as local ones are. Hence, delays in sending immediate relief are much more a result of failure to estimate needs rapidly enough or failure to appeal to the right donors at the right time than they are of the logistical problems of shipping commodities and manpower over long distances (De Ville De Goyet 1993). In the case of a disaster that occurs in a provincial part of a large heterogeneous country, spatial relationships of relief are likely to be governed by a process of mediation, which tends to pit one level of government against another. Thus, local

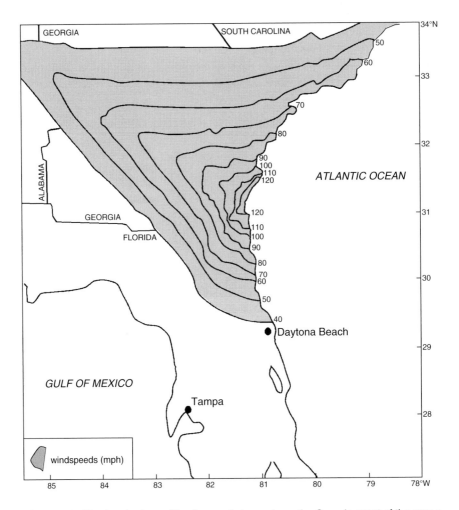

Figure 3.6 Simulated values of hurricane wind speeds on the Georgia coast of the USA: a radial semi-isotropic pattern.

and regional governments press for more resources, whereas national and supranational governments tend to want to limit disbursements in order to preserve fiscal economy. Even in well regulated democracies the process of bargaining can easily become both acrimonious and subject to fits and starts as concessions are alternately made and withheld (Scanlon 1988, Schneider 1992).

The immediate supply of aid to a disaster area is often subject to a core–periphery relationship (De Ville De Goyet & Lechat 1976). Here the dynamics of time are paramount. If monitoring of the impending event is extremely comprehensive, then it may immediately be easy to identify the worst-affected spots in a sudden impact disaster, the core. But this tends to be the exception rather than the rule. Instead, the paramount need is usually to establish the geography of the disaster area, its limits and its areas of greatest destruction and loss of life; in other words, to locate the core but describe the extent of the periphery. At night and in poor weather, that is usually difficult. As a result, the overwhelming preponderance of initial aid tends to be generated locally, and is often crudely improvised, although with the assistance of whatever professional rescue forces are on hand.[37] Initial calls for reinforcements are likely to be unsystematic until order has been imposed on the basis of a general survey of the area. The usual tendency is for outside aid to arrive at the main centres of population and filter slowly to peripheral areas of need (Stephenson 1981, De Ville De Goyet 1993). That is fine when the damage is concentrated in the main cities, but can be highly inefficient if it is distributed across, for example, a group of mountain villages. In such disasters, predictive models and rapid estimation procedures are needed in order to apportion relief more fairly and more rapidly. Such models can only be based on deductive reasoning about modified distance-decay relationships between impact and response, as well as on the pattern of spatial accessibility to damaged settlements.

Organizationally, the key to resolving this problem lies in the tension between centralization and devolution of responsibility. The traditional model by which disasters are managed is a centralist one with a pyramidal hierarchical structure based on a rigid chain of command and headed by a single authoritative figure. This mitigates in favour of a core–periphery relationship in the supply of initial relief, not least because the main command centre is almost certain to be located in the most accessible place, which is also probably the most populous and central. The alternative is an incident command system (ICS; Irwin 1989, Sylves 1991) structure in which there is much less central control. Emergency forces are divided into modular units that work independently under their own smaller command structures, but that cooperate closely with other such units. The key to success lies in the free and rapid flow of information,

37. See Wyllie & Filson (1989) for examples.

which ensures that efforts are not duplicated and tasks do not remain undone. Yet there has been little work on the impact of disaster management by incident command structures on the equity of relief provided over large heterogeneous areas. One can only add that, despite many recent advances in knowledge and training, mistakes are still commonly made, much as in the past (Lord Judd of Portsea 1992, Stockton 1998).

The occurrence of a disaster sends a signal to the body politic and the response to this is often one of legislation in favour of increased public security, hence the concept of a "window of opportunity" in which public pressure on legislators demands improvement (Solecki & Michaels 1994). For example, many of the basic seismic safety laws in California were enacted in response to the patterns of damage caused by the 1933 Long Beach earthquake, especially to schools, which were particularly at risk (Alesch & Petak 1986). Indeed, the repetitiveness of disasters often gives rise to a trail of enactments in favour of stricter norms and building codes, and more highly organized civil protection, a process of improving safety that is therefore stepwise rather than smooth (cf. Fig. 3.4). The question of rebuilding then divides into one of finance and one of realization.

Once again there has so far been little attempt to translate these factors into the terms of deductive spatial modelling. However, a few general observations can be made.

Elements of a deductive spatial model of disaster

Natural disasters can act as an economic brake on regional economies, although they are unlikely to cause economic collapse unless the preconditions for that were already long present (Vinson 1977, Albala-Bertrand 1993). Small and short-lived economic booms may occur, especially in industries that are well connected with the reconstruction effort.[38] Economic health immediately after a disaster is often critically dependent on the way welfare is substituted for market functions, in that gluts of supplies can depress prices, distort markets and create black-market conditions (Winchester 1979). Against this background, reconstruction has to be funded.[39]

In many instances a core–periphery model is once again appropriate at the macroscopic scale. Marginalized regions are unlikely to recover quickly from disaster and will probably have insufficient access to funds and credit, and insufficient source of expertise. However, mere peripheral location within a

---

38. For example, see Geipel (1990).
39. Kates & Pijawka (1977) provided the fundamental model of how this occurs.

country is not a good indication of marginal status: international economic cores can be more important than national ones, whereas marginalized lacunae exist in many highly central inner cities.[40]

At the more localized scale it is not uncommon for recovery and reconstruction to be spatially heterogeneous in a manner that defies simple deductive prediction.[41] The level and rate of recovery can depend on political connectedness, ability to plan and strategize, and the degree of voluntary involvement on the part of outside institutions, from charities to government bodies and investment companies. As a result, it is not uncommon to find stagnation side by side with rapid reconstruction. Legal problems over planning, and difficulty in getting landowners to agree upon collective strategies, can lead to moratoria on reconstruction efforts, especially where the cadastral pattern of property ownership is intricate (Ventura 1984, Alexander 1989d).

Within individual settlements there are slightly more grounds for developing a predictive model. Modern life has yielded a series of innovations that make increasing demands for space, especially in the realm of private transport. Some of the most spectacular of such transformations can be seen in the arrival of mass car ownership in settlements of medieval origin, where living space has traditionally been restricted by the need to defend the population, or through the desire for social conviviality and propinquity.[42] Modern forms of work, leisure, entertainment and social activities tend to be more isolating and to demand at once greater privacy and more space. Rising expectations of lifestyles induce people to search for larger properties, to travel farther and to engage in more diverse activities. The result is that post-disaster reconstruction tends to take place – at least in developed societies – at lower densities than what it replaces (Kates & Pijawka 1977). This can occur in one of two ways: substitution or spillover. In the first instance, there is ample scope to rebuild *in situ* to modern standards. Garages, ramps, lifts, fire exits and so on are added, rooms are enlarged or redesigned, or entire buildings and facilities are substituted by something that is more functionally modern. In the second instance, questions of architectural heritage require that damaged buildings be laboriously reconstructed as they were before the disaster, or alternatively the hazardousness of the site

---

40. In addition, Uitto (1998) and Wisner (1998) discussed the problem of marginalization of individuals in a wealthy city, Tokyo, and how such people may be especially vulnerable to disaster.

41. Hogg (1980) described an example of this in the Friuli Region of northeast Italy after the earthquakes that occurred there in 1976.

42. Increases in car ownership have led to greater vulnerability to disasters (see Ch. 3). For example, the number of cars in central Florence, Italy, is estimated to be about 10 000, many more than there were at the time of the devastating 1966 floods. If flooding were to occur now to the same extent, road congestion would be much worse than it was in 1966, which would offset many of the gains achieved over the past 30 years in emergency response.

places restrictions on what can be rebuilt there. Reconstruction thus spills over to the periphery of the settlement, often on a greenfield site. This can sometimes result in an inversion of urban functions, a suburbanization process, in which central services are transferred to more extensive peripheral sites, leaving the centre with a smaller range of specialized functions or merely a monumental role.[43] This of course, is not a process that is exclusive to disaster but is also a response to more steady changes in the pattern of life. However, disasters can both force such changes and dramatically accelerate them.

Finally, a purely notional geography of disaster exists in the form of hazard perception, a field that has been well studied for the past half century (Parker & Harding 1979, Whyte 1986, Larsson & Enander 1997), and which will be considered again on pp. 76–82. The human psyche tends to magnify the importance of events that are close in time and space, and to diminish that of the distant ones (Turner 1976, Lindell 1994). Fading memory lends a distance-decay function to the temporal pattern of disaster, and physical isolation reduces the significance of events that occur in distant places, unless, that is, the perceiver has had much personal experience of the locales in question (Alexander 1995a). Indeed, personal involvement and experience are the key to the perception of disasters. Of course, it should not be forgotten that, the larger the event, the more attention it attracts. Hence, the dimensions of perception are magnitude of physical impact, intensity of human reaction, degree of personal experience, and level of personal involvement, tempered by innumerable factors of personality, which make individuals more or less sensitive to, or tolerant of, catastrophe (Simpson-Housley & Bradshaw 1978, Aptekar 1991). This implies that an infinite variety of perceptions of disaster exists, which is true, but it would be wrong to suppose that there are no regularities. Sociologists, for example, have identified the "normalcy bias" which induces people to believe information that is more comforting than realistic, assuming, that is, that a choice of interpretations exists (Anderson 1967). Some individuals live by the "gambler's fallacy", which leads them to believe that a recent occurrence of disaster is unlikely to be repeated in the near future, whatever the objective probabilities may be (Alexander 1993: 573). Others suffer "cognitive dissonance", the psychological problem of living with risk but not fully admitting its significance to one's life. In extreme cases this can lead to outright denial of the existence of a hazard threat.[44]

Although the intensity of perception falls off with time and distance, very frequent impacts usually do not generate optimum perceptions, as, when they are small and frequent, the events may be taken for granted. Instead, there is a theoretical frequency and distance at which disasters are perceived most

---

43. In an earlier work I provided a model and field examples of this (Alexander 1984).
44. For example, see Hansen & Condon (1989)

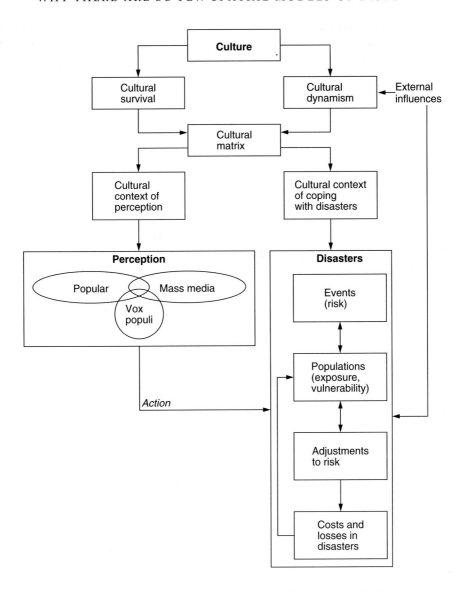

Figure 3.7 The role of culture, context and perception in coping with disaster.

accurately. It may increase slightly and decline more slowly with events whose impacts are particularly large. In geographical terms, perception may attenuate with distance at different rates according to both the magnitude of the event (or its significance) and the perceiver's level of experience of it (Figs 3.7, 3.8).

Hitherto, it has been assumed that disasters are perceived at any distance away in a uniform manner, but this is seldom the case. Various studies have shown that

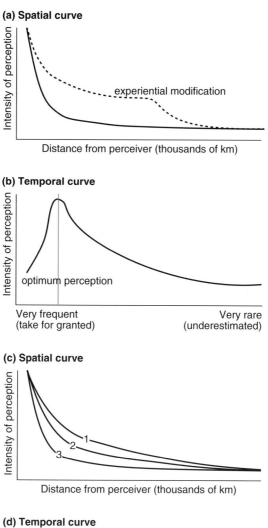

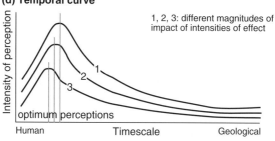

Figure 3.8 Theoretical models of intensity of perception in relation to distance from source of disaster and timescale (showing effect of experiential modification) and time (showing optimum level of perception). After Alexander (1995a).

perception is selective according to what the perceiver finds interesting or important about the event.[45] Hurricanes or earthquakes that occur in popular tourist
destinations may be read about in the foreign sections of the newspapers and
considered more in the light of their impact on vacations than in terms of the
human misery that they have caused. Prejudice is easily fuelled by incomplete
information or ignorance of foreign places, and political predilections can colour
people's views of disaster impacts in particular locations (Alexander 1980, 1995d).
As journalists are mere human beings, they too are subject to such vagaries and
it may show in their reporting and hence reinforce public prejudice. The phenomenon acts as a **perceptual truncator**, which attenuates the extent, the intensity
and the duration of involvement with disasters in far-flung places.

Generally, perception is more holistic and balanced with respect to disasters
that are immediately or closely at hand, and is much more selective as to different
attributes when the events have occurred far away. For example, technological
prowess and social cohesion are recognized in Japan, poverty and overpopula-
tion in India, but these are not necessarily keys to understanding disasters in
those two countries, neither are they absolute and unquestionable tenets. The
split in perception is greatest between foreign and domestic disasters, with the
degree of "foreignness" a rough guide to how wide the gulf is (Needham 1986).

In synthesis, little exists in the way of a secure geographical basis for modelling disasters deductively. It is a challenge that must be faced, and tackled in
three stages. First, robust models of the spatio-temporal distribution of physical
impact and human reaction need to be created. Secondly, they must be cast in
terms of processes, including stochastic ones, that enable them to be realistic.
Finally, there must be a fusion of impact, response and perception models. The
task is worthwhile, because it is a good way to improve our capability to predict
the impact of disasters and respond to them.

Spatial modelling is one task of the geographer, although not his or her exclusive preserve. Another field that has traditionally had links with geography
(again without any proprietorial sense) and that is extremely pertinent to natural disasters is human ecology.

The human ecology of disaster

A preoccupation with the structure and functioning of the natural world led to
the founding of ecology, a discipline based on the taxonomic work of biologists
and pedologists, which examines how organisms relate to each other in the web

---

45. For example, see the classic study of drought perception by Saarinen (1966: 68–9).

of life (Hagen 1992). Whether human beings are merely another living organism or are a species set apart from the others is a rich philosophical question that is probably cannot yield a definitive answer. Nevertheless, it was not long before ecology spawned a branch known as *human ecology* (Whyte 1986). In this, one looks for "the determinants of human behaviour in the natural environment and the processes that facilitate human adjustment to the physical world through social organization" (Mileti 1980). The history of this quest has involved an oscillation between determinism and independence (via possibilism and probabilism), which is not dissimilar in kind to that which has taken place between catastrophism and uniformitarianism. In short, the extent to which the environment determines human characteristics is a matter of interpretation.

Pure determinism involves a dictatorial attitude and one that appeals to people who feel they must have a clear-cut solution to the puzzles of the human condition. Given the similarity of attitudes, it is surprising that the heyday of determinism as an intellectual movement did not coincide with that of fascism and nazism: it occurred about a quarter of a century earlier, although the fascists were quick to revive it (Holt-Jensen 1988). Some laborious scientific work was required in order to demonstrate that equatorial people are not lazy and limited in intelligence because of climatic conditions, and mountain people are not more vigorous and canny because they are continually braced by the weather. Even latterly, despite the demise of the concept of race (which is too difficult to define objectively), there are plenty of recidivists who would resurrect the old discredited theories.[46]

Once physical determinism was formally abandoned, there was a period of prolonged uncertainty, in which the degree to which human characteristics (behavioural ones for the most part) were considered to be controlled by environmental factors vacillated according to what, and who, was studied (Pepper 1984, 1996). Broadly, there was a drift towards liberating humankind from the constraints of habitat. Once again, it was out of step with the political movements, for Josef Stalin had already set *Homo sovieticus* free to pillage and mismanage the environment in the 1930s (Weiner 1988), decades before the accumulated weight of labour-saving technology had created the Le Corbusien illusion of people as being morally superior to the land on which they depended for sustenance.[47] Nature thus took its revenge in the addition of environmental despoliation to the sum of human misery.

---

46. In this, environmental determinism bears a sad resemblance to geological creationism (Shea 1983).
47. In an interesting article, Wang (1981) suggested that, by promoting the "soft storey" or open ground floor that supports the building above on structural columns, called *piloti*, Le Corbusier was indirectly responsible for a certain amount of the vulnerability of buildings to seismic shocks.

There is an interesting codicil to the determinism–independence question. Looked at in another way, it can be considered as the duality of endogenous and exogenous forces, or the impress of environment upon our decalogue as it forms. Are our characteristics inherited or acquired? Is it nature or nurture that forms us? After years of firm belief in the fundamental role of the physical and social environment upon human characteristics and behaviour, the rise of genetics has changed the emphasis quite fundamentally. The current craze to seek, and often speciously find, the causes of so many human characteristics in our genetic code will either reveal us to be much more like programmed machines than we ever imagined we could be, or will eventually have to be supplanted by another revolution in favour of environmental influences. At present it bears a depressing resemblance to the determinism of the previous *fin de siècle*, which was not only misguided but was also used as a moral instrument to justify the impositions of colonialism. The human ecological perspective is now both more diverse and more fraught with controversy than it has ever been in the past. Pioneers, visionaries, recidivists and revisionists vie for public attention. It is all very characteristic of the millennial angst that has gripped the world in the face of nothing more than an abstract way of measuring time and an inability to master humanity's own technological creations.

The field of natural hazards is peculiarly adapted to a human ecological approach, for the simple reason that environmental extremes are powerful enough to exert a strong and consistent influence upon social and cultural systems, if not also on physiological ones (Butzer 1990). This was first recognized in the early 1920s by the geographer Harlan Barrows (Barrows 1923) and was taken up again with vigour in the 1960s by his pupil Gilbert Fowler White (Burton et al. 1968, Burton et al. 1993). By making a series of collective efforts to reduce the negative impact of natural hazards, human societies adjust to them. By employing knowledge, technology and social processes, they adapt themselves to survive and perform under the threat and impact of recurrent disaster. The key to the human ecological approach is the repetitiveness of catastrophe in each place that it threatens. Compared with ordinary daily life, disasters are *non-routine* problems, as the sociologist Thomas E. Drabek has pointed out (Drabek 1989). But in the longer-term perspective they are indeed *routine*, a predictable and expected part of life. If this fact is ignored, then adjustment and adaptation to disasters will largely be innate processes, and very inefficient ones, whereas if it is faced up to there is every opportunity to mitigate the worst of the impacts (White 1974, Whittow 1987).

The recurrence of disasters can be considered in terms of what the geographers Kenneth Hewitt and Ian Burton termed "the hazardousness of place" (Hewitt & Burton 1971; Alexander 1993: 10–11; Hewitt 1997: 164–5). With adequate attention and investment of resources, this is usually perfectly possible to estimate. Although people are not attracted to hazardous areas because of the risks, the latter are often associated with substantial benefits and so they are attracted just the same. Hence, high population levels often exist in fertile farming areas on the slopes of active volcanoes and on coasts with substantial climatic and economic amenity levels but frequent hurricanes and storms. In short, hazardousness does not seem to detract from a place when it comes to settling. Moreover, a variety of motivations bind people to places: economic opportunities, lack of adequate alternatives, *genius loci*, a sense of belonging, and so on. It adds up to a remarkable tenacity in human settlement, a characteristic that requires some further analysis.[48]

As noted in the previous sections, in the history of modern settlement there has usually been progression from cohabitation with hazards without the benefit of risk mitigation, through reliance on increasingly sophisticated forms of structural mitigation, to mixed structural and non-structural methods, with particular emphasis on planning and organizing society to reduce risk. In terms of land use, there are thus four states. In the first, the hazard zone is persistently occupied despite the presence of danger. This may occur without any particular measures to protect the population (as is frequently the case in marginalized communities of the Third World),[49] with no measures other than warning and evacuation (where these are possible) or with a range of comprehensive measures for hazard mitigation and abatement. The state of **maximum geographical inertia** (Alexander 1993: 5) is reached when persistent damage to human settlements does not provoke their abandonment or relocation and the inhabitants therefore cohabit with the damage and rebuild or repair it wherever possible. A form of **secondary geographical inertia** occurs where damage leads to displacement from the original location, but not from the settlement or area in which it occurs. The fourth state – planned or unplanned migration to safer zones – is very much rarer, as natural hazards cause people to abandon settlements and land relatively rarely (although desertification and chronic environmental degradation may be an exception to this rule).

Geographical inertia is a pervasive phenomenon and one that in the Old World is intimately linked to ideas of identification with the *genius loci* of places. In the small town in Basilicata Region (central-southern Italy), where I lived for some years, there was a generously proportioned three-storey façade, behind which crouched a much smaller two-storey palazzo. One of its neighbours lacked a corner and several balconies. These were the effects of earthquakes, but no-one could tell whether they came from the tremors of 1851, 1688 or even 1456,

---

48. In a classic example of this, Burton & Kates (1964b) compared the hazards of inland and coastal flooding. See also Alexander (1989d).
49. See Odemerho (1993) for an example.

so old were the buildings and the damage (Rendell & Alexander 1985). Another town (Senerchia in Campania Region) has been severely damaged by earthquake 11 times in 550 years, but has been neither abandoned nor transferred (Alexander & Coppola 1989). It has instead become another fascinating palimpsest of historical calamities, in which the mute stones tell an enigmatic tale of suffering and partial recovery from disaster amid the iron clamps of the grinding poverty that reigned in the past.

In 1693 the medieval town of Noto, which perched on the shoulder of a deep limestone ravine in eastern Sicily, was decisively ruined by earthquake. The survivors rebuilt it on neighbouring plains and the initiative became famous, not merely for the extraordinary grace and elegance of the new city's baroque churches and palazzi but also for the fact that such a wholesale transfer of urban *civitas*[50] was considered spectacular in its own right, even if the physical distance was only a matter of 15 km (Tobriner 1982). In 1996, one of the columns supporting the drum and dome of Noto's mother church buckled and the superstructure crashed through the roof of the nave and dashed itself to fragments on the floor beneath. This engineering disaster was primarily the result of poor maintenance and failure to monitor developing structural faults, but it could not have been helped by the continuing seismicity of the area, including a damaging earthquake six years earlier. The original reasons for the abandonment of Noto Antica had not been effaced. With regard to the damaged cathedral, the only morally acceptable solution is to rebuild it, where it is and as closely as possible to what it was.

The historical urban landscape in disaster areas is often extraordinarily instructive in terms of what it tells us about people's determination to adapt to hazards (see Fig. 3.9 for some striking examples). While walking down a street in Visso (in the Sibilline Mountains of central Italy), I noticed that the pointed-arch medieval doorways were only half the height that they are in any other Umbrian town. Since the Middle Ages a mudflow had coursed down the street and filled it up with debris. This had not been excavated but merely paved (Fig. 3.9d). In a further example at Calitri, southern Italy, rotational slump landslides had left the back doors of some houses a good 5 m above the street level, but the buildings had not been demolished or even abandoned (Fig. 3.9a).

What accounts for such tenacity? Some reasons are banal, some are fantastic. Property that is ruined or land that is damaged by hazards may lose its value and be difficult to transform into something more productive, especially in places that are economically backward. Although devastation is an eyesore, there often is little economic incentive to clear it away. There are of course exceptions to this. I recall the landslide that opened an ominous curvilinear crack in

---

50. Urban functions and a sense of urban place.

Figure 3.9 Some examples of urban adaptation to landslides in Italy. (a) Calitri, Province of Avellino: headscarp cemented with gunnite; the houses are still used but are not permanently occupied. (b) Bisaccia, Province of Avellino: active landslide headscarp runs under houses, which are still partially occupied. (c) Muro Lucano, Province of Potenza: landslide area in the midst of a high-density urban fabric (identifiable by patch of vegetation in the middle of the town).

Figure 3.9 continued (d) Visso, Province of Macerata: mudflow has raised the level of the street relative to the entrances to buildings flanking it. (e) Craco, Province of Matera: deformation of arches by slow landsliding under foundations of building, which is still in use.

the main piazza of Sirolo, Marche Region (central Italy) in the mid-1980s. For fear of the impact on tourism, the mayor had the crack asphalted over within 24 hours. In the end, the constant threat of disaster is like the fickle attraction of modernity, in the sense that both can lead to the destruction of favourite landmarks. But much is likely to survive, for damage can be rebuilt and modernity may obscure *genius loci*, but it does not devalue it (Morherg-Schulz 1980).

CHAPTER 4

Society and culture

In this chapter we will consider how extreme natural phenomena interact with human cultures and societies. The previous chapter examined some contemporary interpretations of hazard and disaster and a few of the developing trends in analytical methods used by students of this field. This was intended to provide a basis for understanding the social, technological, moral and practical issues that combine to shape natural disasters as phenomena of the modern world. We now move on to consider how extreme natural phenomena interact with human cultures and societies. This will require a blending of themes that are historical, contemporary and predictive; in other words, how we arrived at the current situation, where we are now, and where we are going. To begin with, it is useful to define some parameters for further analysis. One of the most important – and often underrated – of these is the concept of **culture**, which, as will be shown in the next section, has a distinct bearing on how disasters are interpreted and faced up to.

Fuzzy boundaries: disasters and human cultures

In social anthropology, human ecology is more properly known as *cultural* ecology (Butzer 1989). The difference is not a question of transforming the concept but instead indicates that culture is a fundamental variable in the way that people interact with, and react to, their environments. The word "culture" is so all-embracing that it is very hard to define. In essence it consists of the summation of beliefs and behavioural patterns, the imprint of history and the force of achievements of a particular people. It is made explicit in artefacts and symbols, ideas and systems of values. The resulting cultural systems are both the fruit of past actions and a strong conditioner of future ones (Alexander 1991b: 60–61).

This is all very well, but it has to be borne in mind, first, that the individual is usually conditioned by many more influences than merely his or her culture of origin. Secondly, individuals can migrate between and absorb cultures, thus becoming more cosmopolitan with experience and perhaps diluting the cultures they pass through. Thirdly, cultures are dynamic phenomena that can mutate, sometimes rapidly and, finally, the term is not necessarily all embracing. Thus, hazards researchers write of "disaster cultures" or "disaster subcultures",[1] in which the repeated threat or impact of catastrophe engenders a particular reaction among groups that may or may not be culturally defined in other respects; in other words, disaster creates a culture of its own that cuts across pre-existing cultural boundaries. Very positive forms of disaster culture may hold civil protection and hazard mitigation in high regard. Negative ones may perpetuate myths and erroneous assumptions about disaster. Most forms combine both traits. Norms, values, beliefs, knowledge and technology all take on specific subcultural forms (Hannigan & Kueneman 1978, Granot 1996). For instance, in Western societies, which tend to be overconditioned by the electronic media, disasters may be transformed by television news services into a form of voyeuristic entertainment, especially where there is little sense that the viewers are in some way personally involved in the tragedy. It has also been argued (Drabek 1986: 340) that disaster subcultures abate the sense of threat in hazards and can even lead to complacency when coupled with mitigation measures such as warning systems.

Let us return to the question of culture as a universal guiding force. If there are enough elements in the organization and self-expression of a large group of people to permit the definition of a culture, this will contain elements that are both unique to that culture and others that are common between cultures. Researchers have analyzed these in terms of **emics**, the search for the unique or specific, and **etics**, the study of general and universal characteristics (Berry 1969, Brislin 1980, Gherardi 1998).[2] At the risk of oversimplifying, one can hypothesize that cultural survival is a form of **emic value** and cultural dynamism is a form of **etic value**; the former includes inherited attitudes and traditions, the latter enables new ones to be grafted on and the resultant cultural matrix to adapt to change.

But the cultural matrix must again be subdivided. The unique and general characteristics, the ancient and modern, the emics and etics of culture, generate responses in terms of both *perception* and *action* (see Fig. 3.7). The former involves cognizance by the general public, the mass media and officialdom. Together with the latter it enables the signal given by the threat or impact of disaster to be translated into adjustment, mitigation and loss-absorbing actions (Parker & Harding 1979, Whittow 1987).

---

1. Dynes (1970: 79) defined *disaster culture* as "the blueprint for individual behaviour before, during and after disaster". Disaster subcultures, instead, are seen as "group-level coping mechanisms" (Hannigan & Kueneman 1978: 130). See also Granot (1996).
2. A hazards perspective has been provided by Sorensen & White (1980).

The importance of perception to action cannot be overestimated, and the connections made between them are usually based on the idea that people respond to stress in two stages. First, they seek to establish equilibrium by using resources to cope with dangerous hazards. If this fails, they attempt to safeguard vital operations that are needed for survival (Mitchell 1984: 46). Several other dualities are encountered when hazard perception is related to actions taken by the perceivers. First, perception can be divided into that pertaining to the likelihood of damage and that relating to the role of mitigation (Mileti 1980). Furthermore, accuracy of hazard perception is tempered by both the extent to which the resources of the threatened place are needed and the social problems of the population at risk (Burton et al. 1978: 102). In this respect there is considerable variability in the extent to which natural hazard risks are perceived. Some resource users will carry out what seems to be the most appropriate strategy while examining a wider range of alternatives (Haque 1988: 434), but information itself is not necessarily the key to mitigation, even though much hazard abatement is practised in the aftermath of major impacts when information flows are at their most vigorous (Saarinen 1982, Cate 1995). Mitigation action is stimulated more by experience than by anticipation of risk and, on the whole, experience *does* tend to lead to better preparedness (Bollens et al. 1988: 313).

On this basis it is now possible to relate the concept of culture to a fundamental dialectic in human experience, and to consider the result in terms of how we react to disasters in human ecological terms.

Since time immemorial, but especially since the start of the Industrial Revolution, humanity has been faced with a choice of whether to live by exploiting or sustaining the natural environment. The former suggests *parasitism* and the latter *symbiosis*, but in reality there is no such clear-cut distinction. Some degree of artificial manipulation of environmental forces and commodities has always been practised and is a necessary safeguard against many risks to our wellbeing. Moreover, symbiosis is not always either possible or desirable, if it cannot be achieved without major sacrifice. In fact, the consumption of non-renewable resources is not necessarily a "sin" in its own right, and obviously not all pollution is catastrophic or irreparable. There are dangers in associating hazard mitigation exclusively with "green" environmental philosophies, and therefore laissez-faire attitudes to disaster with exploitation of the environment. This is often the case, but it is simply wrong to be categorically opposed to technology when one of its uses is to make our world safer (Pepper 1996).

Nevertheless, technology is a double-edged sword: it can be a source of either mitigation or vulnerability, depending on how it passes through the cultural filter that we use to perceive and interpret it (see Fig. 3.1). A strongly mechanistic attitude to technology can be labelled **technocentrism**, whereas one that emphasizes the role of culture in interpreting and guiding it may be regarded as

sociocentrism. These are not the exact counterparts of parasitism and symbiosis. The reason for this lies in the pluralism of human culture; mechanical and technical ingenuity, and organizational or social ingenuity, are two separate issues.

Yet what we now regard as technology – everything from gadgetry to machinery, plus all the impedimenta of attitudes and procedures that go with it – represents an etic or universalizing input to culture (Brislin 1980). It is translated to the emic aspects through the cultural filter. Good examples of this exist in the means by which rural cultures adapt donated high-technology post-disaster shelters to the traditional needs of their daily lives. This may involve transformations that are startling, original and at variance with what the designer originally intended of the shelters in question (Oliver-Smith 1991); for example, by cutting up donated plastic shelters to make more acceptable forms of sunshade, as was done in Nicaragua after the 1972 earthquake (Davis 1978).

After centuries in which the emics of human societies resisted the etics and predominated over them (times of revolution excluded), we now find ourselves in a period of technologically driven change that has begun to redress the balance. Technological developments have initiated the most drastic metamorphosis of society for a very long time, largely by the way in which they have facilitated transportation and communication. The result is a modern universal consumer culture, one that both facilitates technological development and is incubated by it (Fig. 4.1). Two of its salient characteristics are the tendency to treat disasters as a media spectacle and to view them in terms of accelerated consumption of resources, with all the attendant competition that goes with scarcity.

All of this occurs against the background of cultural survivals, which through the cultural filter interpret and transform the positive and negative outputs of technology. At the moment it is too early to foresee the end of the process. Emic cultural survival is everywhere under threat, but that does not mean that cultural roots which have taken centuries to put down can be eradicated in a few decades, nor are the processes by which society is being metamorphosed necessarily going to do that (Form & Loomis 1956, Alexander 1989d). Prescriptively, the key to disaster management is to learn how to interpret and respect emic aspects of culture and thus to use them creatively to enhance safety, while encouraging the growth of those etic aspects that are compatible with mitigation (Roth 1970).

Traditional and modern cultures

The foregoing discussion has been restricted to several specific dichotomies with cultural and human ecological meanings. The next few paragraphs will offer a more general assessment of the relationship between natural disasters and the human cultures that they impinge upon.

For most people, life is made up of a mixture of ancient and modern cultures. We tend to consider the former as composed of distinctive elements that are unique to the society in question. This may be only part of the story. In reality, even ancient cultures have their universal aspects – the quest for food and security, the designation of places of hegemony, the need for artistic expression. Whatever the balance of uniqueness and generality in ancient cultures, there is no denying the rise and proliferation of modern universalist culture, with its bleak standardizations. It is fuelled by the inexorable spread of commercialism and all the attendant pressures of consumerism and alienation. But to what extent are ancient regional cultures truly in conflict with global commercialization? Are they really in danger of being expunged, sucked dry, trivialized or simply made redundant by commercially based standardized cultures?

Certainly, ancient cultures tend to react to the heavy onslaughts of modernism.[3] One has the impression that modernism cannot be refused. Its allure is too great, or its commercial basis is too solid, or its demands are too insistent. Ancient cultures thus react (Fig. 4.1) by retreating from modernism or metamorphosing

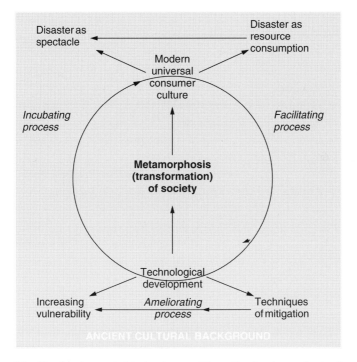

Figure 4.1   The development of technology and the cultural metamorphosis of society.

3. By "modernism" I mean the universal consumer culture, with its heavy mechanistic reliance on technology, standardization and mass production of both goods and ideas, its emphasis on entertainment values, and its economic basis of unfettered capitalism.

with it. In metamorphosis the attributes of historical culture undergo a trans-formation to something that closely reflects the buy-and-sell corporate culture of the late twentieth century. All too easily, beliefs once deeply held become products to sell, or at least to demonstrate to outsiders.[4] Regional specialities and ancient traditions are reduced to the level of mere sustainers of the local economy (perhaps a necessary act, but culturally a ruinous one). By contrast, the retreat of local culture tends to make it more subtle and elusive, and less acces-sible to outsiders. Its very protagonists may be unaware of the origins and cul-tural significance of their attitudes and of how these have been forced into retreat by the relentless onslaught of modernism. Consumer culture takes no prisoners: if it cannot transform the ancient cultural mores, it will belittle them and usurp them until they dwindle away.[5] The process is illustrated in the Palio horse race, which for several centuries has been the obsession of the central Italian city of Siena. In the city's culturally distinctive neighbourhoods, the *contrade*, the semi-annual race, which lasts less than three minutes each time, is lived every day of the year with an intensity that outsiders find incomprehensible. But television's intrusive eye has striven to turn the Palio into a pseudo-medieval equivalent of Formula One motor racing. Its significance as viewed between the commercials and the evening news is vastly less than what the *contrade* get out of it. But television brings money, advertising, interlopers, debate, controversy (anti-vivisection leagues that protest about cruelty to racehorses) and a variety of more or less subtle influences that tug at the cultural roots of this curious old tradition. It is hard to say that the Palio is enriched by mass participation; such is the impov-erishment of humanity in the age of global copiousness.

Left to their own devices, traditional cultures absorb and resist the suffering caused by disaster impacts. The degree of mitigation that they give rise to is highly varied but generally low. Protection against disasters tends to take the following forms: the wise use of land, the painstaking creation of surpluses, and the evolution of safer forms of building. But none of these is universal. In terms of the first of these, the farmers in the delta of Bangladesh and the stone-age cul-tures of the high Andes both developed a highly sophisticated relationship with the land they cultivated and used modest but painstaking means to control the hazards they faced (flood and drought in the first case, landslides in the second; Oliver-Smith 1977b, Shahabuddin & Mistelman 1986). In the third case, in the Caribbean vernacular housing evolved (Snarr & Brown 1994) that would resist hurricane force winds and, in the Pattan Valley of Pakistan, that would resist earthquake shaking (Coburn & Spence 1992). Monumental architecture evolved

4. In Fiji, celebrants go into a trance and walk barefoot over incandescently hot stones every Thursday at 7.30 p.m. for the benefit of Australian tourists.
5. Hence, for example, the prevalence of fashion items that are distinctly French but are made in Indonesia.

Figure 4.2 Inca stonework at Sacsayhuaman in Peru. Note that the fit of the stones, assembled without mortar, and the pattern of re-entrants (offset walls) virtually guarantees freedom from earthquake damage.

in Inca Peru (Fig. 4.2) that would resist major earthquakes, in the former case through the close fit of the stonework and in the latter by employing a ring beam to hold the walls together. But what is more remarkable is the lack of an "architectural Darwinism" in history of building. More often than not the structures that fell down or were swept away were reconstructed to the same vulnerable standard. Historically, and therefore culturally, disasters are recurrent events. They have thus been codified in the elaborate systems of lore and beliefs that form the basis of many ancient cultures. Although Sir James G. Frazer may have been right in *The golden bough* to see a broad transition in human thought from magic to religion to science, so much of the earlier phases remains with us, especially in the rich legacy of interpretations of disaster gods to propitiate, rites to perform and signs to interpret (Vitaliano 1973).

Modern culture is more flexible and is both the vehicle and the artefact of technological change (thus it both creates such change and is fostered by it). For all its faults, modernism permits more ways of adjusting and adapting to disaster than does traditional culture.[6] But it also creates many more and many new forms of vulnerability. The elements of modernism can be codified and can be viewed in terms of their relationship with disaster, as follows.

---

6. See Schware (1982) and Pramanik (1995) for examples.

Commerce and consumerism

Many more people participate in disasters than previously, but not necessarily in constructive ways. Although concern and charitable giving have waxed large over the past 30 years,[7] a new phenomenon has arisen: the presentation of tragedy and misery as a media spectacle. Suffering, hardship and misery are sanitized, tidied up and served up for human consumption by the rich and fortunate peoples of the world. This phenomenon has gained the epithet "disaster pornography" (Omaar & de Waal 1992).

Consumerism is also the driving force of the repetitive fashion for disaster movies, which often distort reality, reinforce prejudices and trivialize human relationships by superimposing entertainment values on them. Consider the film *Dante's Peak*, which is fairly representative of the genre. A young volcanologist and the lady mayor of a small town on the flanks of an active volcano observe the signs of an impending eruption and have to struggle against powerful vested interests in science and real estate in order to ensure that their prediction is taken seriously. In the end they are vindicated by the destruction of the town in a massive outpouring of ash and lava. The eruption is made more spectacular by utilizing five different styles of volcanism – the Hawaiian, Icelandic, Strombolian, Vesuvian and Plinian – all of which are characteristic of radically different volcanic settings. Civil protection is dramatized by simulating absolute incompetence, stupidity and utter lack of experience on the part of both the authorities and the leading volcanologists. The usual clichés of civil protection are propagated, including panic in the population, chaos in the provision of aid, and authoritarianism on the part of civil and military relief units. Only personal heroism can save the day (which is never the case in disasters). As is usual in such films, the drama reduces to a struggle for leadership based on black and white concepts of good and bad behaviour. The result is anything but subtle, but it is considered to be good entertainment: death and suffering are turned into a modern saga.

The final aspect of consumerism that is relevant here can be found in the question of profit and loss in disaster areas. Both Africa and the Indian subcontinent have suffered considerably in the past from profiteering in the sale of scarce goods after disaster, whether they be foodstuffs, seedstocks or the raw materials of housing (Downs et al. 1991). Although the international aid donors have proved very sensitive to the problem, it cannot entirely be stamped out. Secondly, in some countries, notably Afghanistan and Colombia, there is an intimate relationship between disasters, armaments and the narcotics trade

---

7. In other words, since the Nigerian civil war of the 1960s first gave Western television viewers an intimate view of the tragic effects of famine (Mayer 1969).

(Olson 1995). Although it is difficult to make valid generalizations, it seems that gun running and drug exporting tend to destabilize such societies and impoverish them, while enriching certain elites. The result is an inability to recover effectively from the frequent disasters that affect such countries: resources are not distributed equitably, democratic processes are not adhered to, repression and violence are common, and it is dangerous to protest. At the same time, outside intervention may be limited to the international community's preoccupation with controlling the distribution systems for armaments and narcotics, rather than ensuring local self-sufficiency in the face of earthquake or flood risks.

Market forces

Economists have long thought of disasters as brief periods distinguished by unusually fast consumption of resources (Jones 1987). Although in theory the rules of the market are suspended during the emergency phase of disaster (and they may be modified or mollified for a long period thereafter), in practice disasters give rise to rapid and somewhat arbitrary shifts of resources. At their most extreme, these amount to attempts to dump surplus or outdated products on the hapless victims of catastrophe.[8] Even where the relief supplies are fully justified by circumstance, much of the altruism has gone out of disaster aid.[9] Donor governments are under pressure to economize. Part of the process, the logic of the market, is to reclassify disaster relief as a form of investment on which a demonstrable return is required.

Global market

The globalization of production has seen its regional differentiation. It is an all-or-nothing situation in which countries or societal groups are either fully fledged participants or are excluded and left behind in the race for material prosperity. In all societies, marginalization is a key element in vulnerability to disasters (Twigg & Bhatt 1998). For the poor and marginalized, the unemployed and the homeless, the strictures of monetarism are anathema. Mitigation and social protection become casualties of programs designed to contain national debt. At the same time there has been a more benign form of globalization: the growth of the international relief network, with its thriving industries of

---

8. This was particularly notable after earthquakes in Guatemala in 1976 (De Ville De Goyet et al. 1976) and Armenia in 1988 (Autier et al. 1990).
9. Indeed, after the boom years of the early 1990s, the international relief community is facing a crisis of both morals and confidence (Hendrickson 1998).

appeals, donations and shipments (Borton 1993).[10] But vulnerability continues to rise, and the gap between the rich and the poor carries on widening. Hence, the growth in international solidarity is unlikely to match the rise in need.

Failure of international relations

International disaster relief has suffered instability and also a moral and operational crisis. In part, this results from the inability of the world community to create an equitable system of international relations in the wake of changes brought about by the fall of the Iron Curtain. Imbalances of power and wealth have deepened considerably: for example, in 1960 the income of the 20 per cent of the world's population that lived in the richest nations was 30 times greater than that of the 20 per cent who lived in the poorest countries, whereas in 1998 it had become 82 times greater. This means that the global, regional and, often, the national stability required to bring lasting relief from the recurrent affliction of disasters has not been created (Duffield 1996).

Accelerating consumption of resources

Rising population and rising expectations lead to a faster rate of consumption of resources. If disasters are indeed periods of sudden heavy resource consumption, then heavier losses are to be expected and, moreover, at a faster rate than hitherto (Berz 1991, Kunreuther & Roth 1998). As new ways to use resources are continually being invented, so new and enhanced sources of vulnerability are bound to emerge. Accelerated consumption also increases the amount of fixed capital goods that are at risk of damage by disasters.

Standardization of tastes

There are good grounds for suggesting that disasters have a disproportionately serious effect upon older and historic property than upon the new. Moreover, while the world's brief attention span is focused on a particular disaster area, there is a greater opportunity to bring in new influences with the relief supplies. Hence, the standardization of taste is accelerated by disaster.

---

10. However, in the later 1990s financial allocations to international relief began to tail off (IFRCRCS 1996, 1997).

The "information society"

Information and knowledge have never been so freely and copiously available than at present. Each day brings new sources and new means of access. But the emphasis is squarely on quantity at the expense of quality. It used be said that "knowledge is power". The current plethora of information has changed that and has incidentally shortened the world's attention span. It may still be true that knowledge of a certain type and quality (i.e. wisdom) is power, but the cheapening and trivializing of information flows, the chopping up of news into palatable chunks and the primacy of the "sound-bite culture" has left the acquisition of wisdom as arduous, difficult and exclusive a process as it ever was before the information revolution. As knowledge is a prime requirement in the mitigation and management of catastrophes, these developments have profound repercussions for disaster prevention and civil protection (Cate 1995).

Information technology

The latest revolution in disaster management is the full application of real-time information technology. But it is a very partial upheaval and its high potential for resolving problems will probably only be realized if it is able to interact in a very sensitive and flexible way with our cultural systems (Quarantelli 1996).

Social isolation, fragmentation and polarization

The mass prosperity offered to some sections of humanity by triumphalist capitalism is achieved only by paying a very high price in terms of social isolation. The motor car, the computer, the video cassette recorder, useful though they all are, contribute to a partial withdrawal from traditional forms of social participation. The so-called "virtual" worlds are not true substitutes for reality, but are a response to the processes of social fragmentation caused by reliance on technology. This demands a new interpretation of the ideas of "disaster culture" and "disaster subculture", in which involvement is simulated through the so-called "virtual reality". Thus, the use of computers in emergencies has created a disaster subculture in its own right. At this point it is time to pose the question "Are disasters events that cause socialization or isolation?" Traditionally and historically the answer has been clear: emergencies give rise to more or less universal processes of social participation (O'Brien & Mileti 1992); long-term aftermaths can easily lead to isolation and abandonment. Perhaps it is time to re-evaluate that conclusion in the light of modernist culture, for it may be that in some cases

71

the sense of social isolation, fragmentation or polarization never really disappears, even during the convergence reaction at the height of the emergency. That which did not exist before the disaster cannot be created by improvization during it, which is a valid general principle for many aspects of catastrophe.

Disasters and social change

Having considered some of the facets of modernism, it is now opportune to examine how they combine to create general processes of social change, and how these are currently affecting attitudes to natural disaster.

At the turn of the century and millennium, the pace of change in everyday life has never been faster. World population has grown more rapidly than the Earth's capacity to accommodate the people who must live off its resources; or, more correctly, population has outstripped the social capacity to distribute Earth's relatively scarce resources equitably.

Technology has developed and proliferated faster than the disadvantaged people and their social systems have been able to assimilate it. In addition, it is macabre that technological development has enabled killing, maiming and torturing to become more efficient and widely distributed than ever before (Macrae & Zwi 1992). For the privileged, expectations have risen to levels far beyond what may be regarded as sustainable living in the long term. Globalization has overrun the localized fixed points of people's lives; attitudes and habits have become standardized. News, entertainment and cultural pursuits have been copied from one culture to another to the point that they have become universally accepted. Easy international communication in real time has facilitated the growth of a global market in which capital has gained the upper hand over labour. The resultant form of international capitalism has changed the nature of government, consumption and consumerism, mass attitudes, employment and many other fundamental aspects of life.

So rapid and so momentous are these changes, so deeply are we involved in the current transition (but transition to what?), that it may be too early to assess their impact on society. Yet one thing is certain: an emergent science of "disasterology" will have to face up to the new parameters of life and human culture. Consider the following chain of reasoning:

1. Disasters result from the impact of extreme geophysical events upon vulnerable social and cultural systems.
2. Without human vulnerability and the assumption of risk there will by definition be no disaster.
3. Vulnerability is an artefact of the physical and economic constraints upon

life, particularly scarcity, but it is shaped by perceptual, social and cultural mores and traditions.

4. Culture shapes the collective and individual reaction to the threat, risk and impact of disasters.

5. Therefore, culture is a fundamental determinant of vulnerability and socio-economic impact of disasters.

If in these times of transition it is too early to see the end of the momentous processes of change that are sweeping the world, it is still possible to ask pertinent questions and seek interim answers. To what extent can new ideas and approaches be grafted onto old ones when disaster strikes, or must the new necessarily supplant the old? Hence, do old cultural matrices resist the impetus of change or do they absorb it? If they absorb change, does culture lose its functionality and *raison d'être* or do the cultures themselves lose these properties? In the light of global change and international disasters, can we distinguish functional from dysfunctional traditional cultures? In effect, does "cultural disease" occur, and does it inhibit adaptation to disaster?

Cultural disease and its effects

Any attempt to answer these questions must begin by returning to the definition of culture, with special reference to the study of disaster. The concept of culture is complex and elusive. Social historians have defined it as "the weaving of values, aspirations, beliefs, myths, and ways of living and acting as they articulate on the level of a [common] mentality" (Maravell 1979).[11] Science has been prominent among the strands that go to make up Western culture and values, although the choice of what to study scientifically and how to study it has been strongly influenced by the sense of values induced by each scientist's "cultural filter", which in cumulative terms amounts to a source of ideology. In turn, ideology can be deterministic, probabilistic or possibilistic in proportion to the degree of certainty or uncertainty in its cultural and historical referents. Indeed, environmental perception can assume such a strong cultural ethos that we end up in a parallel world, the homosphere or noosphere, which is conditioned by thought and action, rather than by unfeeling natural processes (Svoboda 1999, Svoboda & Nabert 1999). Culture is paramount in this. For instance, in natural hazards, decisions about whether to mitigate a natural hazard are often not a function of how dangerous the hazard is in absolute or objective terms but how dangerous it is perceived to be.[12] For example, the eruptions of Mount St

---

11. See also Alexander (1991b).
12. As noted by White (1973).

Helens in the US state of Washington in 1980 killed 65 people, but were small by the standards of explosive volcanism (a maximum of 4 on the volcanic explosivity index, which runs from 1 to 8; see Alexander 1993: 33). Yet the impact on the US mass media was unprecedented. In response to pressure from the public, funding for volcanic hazards assessment was subsequently increased by 1500 per cent (Monmonier 1997: 58). In a second example, my own fieldwork in southern Italy revealed that the principal reason for discrepancies in the degree of earthquake protection in a group of seven villages was the presence in one of the settlements of the house of a cabinet minister's aunt. This village, which was the least at risk, had by far the best seismic mitigation program, and there was little protest at this state of affairs, for local people accepted the discrepancy as a natural outcome of political patronage that had existed there for centuries in fundamentally unchanged form. Thus, collective culture has the ability to create, enhance or ameliorate vulnerability.[13] Such manipulation is achieved by expressing preference through choice, under the three constraints of affordable cost, state of knowledge of the phenomenon under consideration, and the technology available to modify it. In this respect, no culture has been able completely to screen out the egocentric tendencies in human beings.

In reality, "cultural disease" can be identified but it cannot be regarded solely as the affliction of ancient cultures. The symptoms of cultural malady are found in widespread corruption of public office, the rapid growth of mafia and other forms of organized crime, the arbitrary exercise of excessive power, the alienation and marginalization of groups of the population, and the rank exploitation of people who lack the resources to stand up to their oppressors. Reactions vary from the deterioration of public services to forms of low-intensity urban warfare. In all instances these ills reduce society's capacity to mitigate disaster, and they increase vulnerability and concentrate risk in the poorest and most marginalized sectors of society.[14] Sometimes the roots of cultural disease can be traced back to ancient oppressions or longstanding but unresolved conflicts, and at other times the spores are carried on the winds of globalization.

Many of the symptoms of cultural disease can be found in relatively benign form even in the healthiest of societies. The problem becomes significant when they are present in multiple form and at chronic levels, In point of fact, there are two kinds of symptom: **socialized** and **divisive**. The socialized manifestations include organized crime, corruption of officialdom, the abuse of power and excesses of nationalism, separation and the so-called religious fundamentalism. The divisive forms include marginalization, alienation, exploitation and the creation of an underclass. Socialized phenomena tend to absorb and mutate the

---

13. A scientific analysis of this has been provided by Denis (1997).
14. No better example exists than that of the Sudan (Allen et al. 1991).

effects of disaster, but they are also propagated by it. Through failure to connect relief with development, the divisive phenomena are accentuated by catastrophe.

In point of fact, the two forms of malady are not as separate as may appear. For example, marginalization and alienation foster the growth of nationalism and fundamentalism. Thus, the latter are forms of extremism that grow out of social polarization. The source of nationalism is that it is an ancient frustration that has found fertile conditions for resurgence in the new global order as strategic realignments permit or even encourage the re-emergence of separatist movements. Now, even in the most homogeneous of societies, disasters initially create a strong feeling of social identification and solidarity between groups, but they subsequently tend to accentuate the divisions that are endemic in society. Nationalism, separatism, and fundamentalism tend to resist the social cohesion engendered by emergencies. In the worst case (e.g. the Ethiopia–Eritrea wars of the 1980s), disaster relief simply became an instrument of the conflict (Clapham 1991): relief convoys were bombed and strafed, food supplies were hijacked, non-combatants were denied relief, and people were driven off the land and into the towns, where they would not be able help the enemy.

Now we return to the essence of culture and its relationship with disaster. Time is the principal dimension of disaster, as the temporal progression of events is a very strong feature of most catastrophes. It is also instrumental in the transformation of events into history (Alexander 1991a, 1994a). This is both a summative and a selective process; the cyclical nature of events means that history appears to repeat itself, but the unique combinations of process, rate, magnitude and intensity mean that in reality there is no going back and no cyclicity. In any case, history is absorbed into culture, or better, it *fashions* culture by creating the conditions for collective reaction to disaster.

As noted at the beginning of this book, disasters represent milestones in the long march of time. They are fixed points of reference in people's lives, in the evolution of approaches to safety and to civil protection, and in the cycle of destruction and reconstruction. How then is a history of such events transformed into an ingredient of human culture? At what level of society does culture *exist* in relation to its determinants such as disaster? We should note here that regional cultures tend to be more powerful than national ones, and in many cases the latter are artificial constructs.

In order to understand how disasters influence and transform human cultures, it is necessary to consider them in terms of how they are perceived and assimilated by the people who experience them. Hence, the next section will offer some further thoughts on the perception of hazards, risks and disasters, with particular attention to the cultural referents of the perceivers. (See the footnotes in this section for a more systematic account of hazard perception.)

The perception of disaster

When in the 1920s Vladimir Vernadsky and Pierre Teilhard de Chardin formulated the concept of the **noosphere**, they thought of it as a purely conceptual or cognitive realm that interacts with the biosphere, atmosphere, hydrosphere and lithosphere by virtue of humanity's ability to transform thought into action (Fig. 4.3).[15] Environmental perception began to take on a life of its own: it was recognized that the interpretation of nature, the human experience of it, can be as important to our species as nature itself can be, and the two are not necessarily the same (Burton & Kates 1964a, Bates & Pelanda 1994). Natural hazard and disaster perception have gradually been revealed by patient research over the past

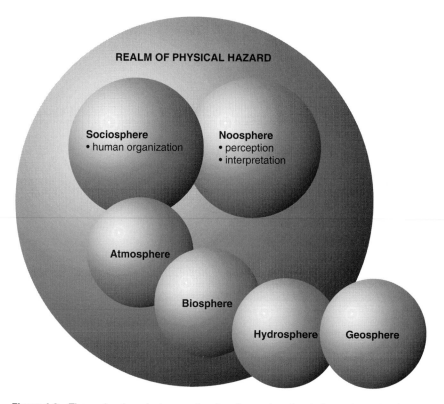

Figure 4.3 The sociosphere (or homosphere) and noosphere in relation to the realm of physical hazard.

15. See Svoboda & Nabert (1999). In a comprehensive review of anthropocentrism in human thought, Meyer-Abich (1997) noted the words of Alexander Von Humboldt (*Kosmos*, vol. I, p. 69, 1845–62): *als wäre das Geistige nicht auch dem Naturganzen enthalten* (that the world should not be conceived, "as if mind were not included in the whole of nature").

50 years on the basis of the human ecological approach to environmental extremes (Mitchell 1984, Whyte 1986, Lindell 1994).

The process of perception is a subtle one. External stimuli received by the subject are organized in order to recognize and know them, or in other words to make sense of them by virtue of some prior ordering system that is held in the mind and that has been acquired on the basis of past education, training and experience. Understanding this has a great deal to do with perception, as do attitudes, ideas and feelings: in short, perception is a partial and selective process. The search for its broad regularities is hampered by the fact that we all have unique backgrounds and histories, but it is aided by the common well of shared experience and the tendency to react in similar ways to particular stimuli. Nevertheless, it can be a highly complex task to identify the characteristics of people that help determine their modes of perceiving hazards (Simpson-Housley & Bradshaw 1978, Denis 1997, Palm 1998). These may include education, socio-economic status, family, marital status, occupational and employment status, religion, cultural and ethnic background, past experience with disasters, and degree of mental stability. However, British and American research has frequently found that hazard perception seems to act independently of the socio-economic variables (Van Ardsol et al. 1964, Parker & Harding 1979). Other researchers have offered more positive results (Rasid et al. 1992). A further difficulty is presented by the "double layer" of perception, not merely that of the subjects, but that of the researcher who analyzes them. To illustrate what is involved in the latter, we may note a fundamental difference of opinion regarding the significance of disasters. Are they "evil incursions" or "black holes" in the peaceful fabric of daily life, as Michael J. Lindell would have it (Lindell 1994), or are they such normal events that they "reinforce the myth of ordinary life". as Kenneth Hewitt has argued (Hewitt 1983)?

In an ideal situation an individual faced with a natural hazard would appraise the threat, study the range of alternative mitigating actions, evaluate the consequences of each particular option and choose the action, or a range of actions, best suited to the circumstances. In reality this tends to be too difficult a strategy for most people, who live in a world in which rationality is bounded by the limits of perception and reasoning power (again possibly linked to the personal characteristics listed above).[16] For instance, people tend to evaluate alternative actions sequentially rather than concurrently, which may limit their ability to make comparisons. Moreover, innate conservatisms and resistance to new ideas may lead many people to distort new evidence about natural hazard risk to conform with their own ingrained perceptions. In this respect a continuum exists in the *locus of control* from *internal control* (belief in one's ability to

---

16. The "bounded rationality" of Simon (1956).

make decisions independently) to *external control*, in which events are believed to be influenced largely by chance, luck, external factors and other people (Whyte 1986: 254–5).

One of the worst situations is the state of "learned helplessness", in which knowledge of the hazard is adequate but it appears that no amount of adjustment will mitigate it and so people stop trying (Hansson et al. 1979: 415). A corollary is the state of "cognitive dissonance", in which an individual perceives his environment to be hazardous but continues to live in it (Preston et al. 1983). This can give rise to outright denial of the existence of the hazard, a trait that so struck earthquake perception researchers in San Francisco and Los Angeles (Turner et al. 1986, Hansen & Condon 1989). In less extreme circumstances the options for adjustment to hazards can be limited enough to define what Robert W. Kates has termed the "prison of experience" (Kates 1962: 140). In any event, the rigorous intellectual activity of making decisions about risky events is beyond the competence of many people, and hence inaccurate perceptions may prevail even among the best educated and best informed members of society.

Nothing fuels perception like the stress of an impending or actual disaster. The first reaction is usually one of trying to cope directly with the risk or impact. If this proves inadequate, people devote themselves to safeguarding the resources they need in order to survive (Mitchell 1984). Greater levels of experience and adaptation do tend to reduce anxiety levels, whereas uncertainty and unfamiliarity increase the stress. Besides this, several coping mechanisms rely on suboptimal perception to protect the individual against the psychological reality of hazards. The "syndrome of personal invulnerability" leads people to suppose that disasters always happen to someone else. The "gambler's fallacy" lets them believe that the occurrence of a particular event in one year will make it less likely to recur soon afterwards (which may be true for some hazards, perhaps earthquakes, but is certainly not so for others such as floods). Weak and inaccurate perceptions result from failure to assimilate or understand hazard information and from the tendency to make incorrect assumptions about risks (Kilpatrick 1957). Lastly, people may put their faith in precautions that are unlikely to function. In short, the negative aspects of hazard and disaster perception seem to be more striking than the positive ones.

Perceived risk can be considered in terms of three factors: dread, familiarity and exposure (Slovic et al. 1979). Low levels of exposure decrease the attention given to hazards, although very high levels may do likewise. The arrival of disaster tends to provide a rush of adaptations to mitigate future impacts (again, the "window of opportunity" for hazard mitigation – see pp. 1–5, 7–22 and 40–53), whereas recent or frequent experience of disaster tends to increase both knowledge and sensitivity levels (Parker & Harding 1979, Alexander 1995a).

Two other factors are vital to an understanding of hazard perception: context and culture. It is a mistake to view the perception of hazards and disasters in isolation from competing demands on people's attention from other events. Both the scale and the accuracy of hazard perception may depend on the extent to which lives and resources are at risk and the context of social problems in the community. Likewise, hazard perception must be viewed in its cultural context (Fig. 3.7). Social relations have a gestalt that extends the influence of culture and history far and deep into the psyche of the individual and the group to which he or she belongs. Thus, collective perception and action result not merely from the sum of individual perceptions and actions, and of the prevailing social milieu, but also of events in history that cast a long and enduring shadow over the present and the future. In the most straightforward cases, chronic and well known threats are assimilated into the local culture and lead to predictable interpretations and reactions. Thus, we end up with a "disaster culture", a model of individual action and reaction in relation to disaster. Its roots may be religious, political, social or, even conceivably, scientific, and it represents a form of consensus in society. Yet as societies are not uniform entities, we may also define "disaster subcultures" as groups that adopt shared perspectives but are not necessarily homogeneous in any other way (Hannigan & Kueneman 1978: 130; Granot 1996). The role of disaster cultures can be decidedly mixed. For example, the 1964 Great Alaskan Earthquake led to changes in the community of Old Harbor (Davis 1986). Political awareness and ethnic identity were enhanced, knowledge of available services improved, but dependence on government agencies also increased, and hence resilience decreased. In the most notorious cases, disaster subcultures are heavily based upon aid dependence (Dirks 1979; Smith 1996: 75).

A critique of research on hazard perception

Having discussed some of the characteristics of hazard perception, it is worth taking the research with a grain of salt and endeavouring to construct a critique of it. To do this, we should note that structural mitigation tends to concentrate on the *consequences*, rather than the *causes*, of hazards. Thus, the increasing reliance on non-structural measures highlights the importance of perception, which is integral with prevention and adaptation.

Nevertheless, perception research has some inherent weaknesses that have so far led to a reluctance to utilize it. To begin with, conclusions are often insufficiently specific to have much practical meaning. Secondly, there has been a tendency to confuse perception with the actions based on it, even though the relationship between them is not linear (see pp. 7–22 and 40–53). For instance, perception of imminent danger may lead to panic and flight, but panic is merely a way of externalizing the perceptual reaction, and a rather controversial one.[17] Thirdly, there have been some startling discrepancies in the perceptions and attitudes of hazard-zone residents, hazard managers and academic researchers, especially in the poorer parts of developing countries. Green et al. (1991: 228) cited evidence from Zambia to suggest that the terminology used by consulting engineers is alien to local cultural traditions, which embody a greater degree of adaptation to flooding and enable residents to perceive floods in a less threatening manner than the engineers assumed. In western Bangladesh, Brammer (1987) found that farmers applied their varied repertoire of techniques to mitigate drought long before the government had begun to take note and organize remedial and relief efforts. Even in wealthy Japan there are discrepancies: researchers had difficulty matching their own conceptions of hazard perception with those of their respondents, who found it difficult to understand Western concepts such as "bad", "luck" and "emotion" (Whyte 1986: 248).

The "prison of experience" and accompanying sense of helplessness increase with the level of poverty and deprivation of respondents. Thus, even in the most developed societies, there are groups of disadvantaged people who tend to look on hazards as an unavoidable affliction, a relatively unimportant matter, and a secondary concern in relation to poverty and social alienation. In many developing countries, the apathy, demoralization and isolation induced by hazards are part and parcel of the condition of poverty. For the marginalized, the real disaster may be perceived to be not the flood or earthquake but the government elite or the local landowner (Blaikie 1985). On occasion it may even be perceived to be the purveyors of relief and development aid (Dirks 1979, Stockton 1998). Among the international relief community a lively debate has arisen concerning the value of folk wisdom and other forms of indigenous knowledge.[18] Although rural populations have often been underestimated or ignored by aid agencies, there is no sense in making a cult out of peasant wisdom at the expense of technology transfer. On the other hand, education has sometimes been viewed as little more than cultural imperialism, as the educated classes may be the exploiters who contribute much to the problem and little to its solution. The other side of the coin is lack of public confidence in governments in which the state is seen by its people as far from benign and perhaps capable of double standards or divide-and-rule tactics. Often, the poorer groups fear extortion, looting and separation from the kinship and neighbourhood groups that lend them sustenance, which may mean more to them than a remote government that does little to protect them from unscrupulous local political bosses (Parr 1970, Haque & Zaman 1989).

---

17. Compare Quarantelli (1954), Keating (1982), Johnson (1987) and Alexander (1995e).
18. For example, Schware (1982), Rasid & Paul (1987), Paul (1999).

Until the 1990s, and perhaps still, the overwhelming preponderance of research on hazard perception was conducted in affluent countries, particularly
the USA. This has inhibited the development of a cross-cultural standard that
would enable greater comparability of research findings. Thus, we have little
basis for making robust universal generalizations about perceptions and attitudes to hazard and disaster. All we can say is that natural hazards are much
less salient in the perceptions of poor people in developing countries than they
are in the minds of the wealthy and of people in developed nations.[19] Emphasis
on personal liberty and degree of choice is considerably greater among the
inhabitants of rich countries than among people at risk in developing nations.

The American cultural viewpoint is ill suited to the role of world model for
perception studies that it has tended to assume. As its crudest, American society
is fiercely mechanistic and obsessed by gadgetry, such that technologically
complex solutions to the world's problems acquire more kudos than simple
ones. It is by nature eclectic, but it is also highly individualistic, isolationist,
antagonistic, expressionist, materialist, and sometimes triumphalist. Hence, the
American approach to hazard assessments tends to assume, first, that people
are driven by free will rather than what J. D. Bernal called "the freedom of necessity" (Bernal 1949) and, secondly, that individuals mainly participate in society
out of desire and choice, rather than outright obligation. These are assumptions
born of a nation forged out of disparate elements, mixtures of idealism and
internal conquest, and constant antagonisms between the business of commercialism and the unquantifiable demands of human relationships. American
society has a collapsed view of history: it is sometimes almost ahistorical and
it easily fails to recognize the subtle veins of historical influence in longer-
established societies elsewhere. For these reasons, any attempt to export the
American view of hazard perception risks being a form of cultural imperialism.

Two other characteristics distinguish American society from the point of
view of perception. The first is its extraordinary social and spatial mobility. The
latter has caused many people to settle in unfamiliar surroundings subject to
risks of which they have little perception or experience (Burton & Kates 1964b,
Birkland 1997b). This reflects heavily on the perception of hazards. The second
is the emphasis that has been placed on *hazard*, rather than *disaster* perception.
The latter implies a more restricted range of choices and a greater degree of fatalism than the former. It points once again to a difference in perception between
First and Third worlds.

Fortunately, an alternative model of hazard perception has begun to emerge,
thanks especially to social scientists working in Bangladesh.[20] Research began in

19. See Zaman (1989) for an example.
20. For example, Paul (1984) and Nurul Alam (1990).

the 1970s under the auspices of the Burton, Kates and White approach as developed in North America (Islam 1974, White 1974), but the conclusions soon began to diverge under the impetus of a very different social reality. Paradoxically, although the emphasis is on perception of constraints rather than opportunities, research findings have begun to constitute an opportunity for change in their own right, through giving voice to the communities that are directly involved with natural hazard risk (Westcoat et al. 1992, Paul 1997).

Nevertheless, we are a long way from a theory of hazard and disaster perception that unifies human experience across diverse cultures: the "etics" of the field have not been properly established. In this respect, hazard and disaster perception have usually been analyzed with insufficient attention to their context. I contend that they should be treated as an extrapolation of people's perception of less extreme phenomena and events. These can be very revealing of the social, cultural and technological impress upon people's attitudes, and these influences may intensify, rather than diminish, in extreme situations.

Having considered the role of perception and culture in the human response to disasters, we now turn to another problem that has become salient in this millennial world of striking contrasts: abundance and shortage. The polarization of resources, the widening of the poverty gap, and the consolidation of economic and political power have tended to accentuate the impact of disasters. It is a curious paradox that abundance continues to co-exist so firmly with deprivation, and the situation has worsened in recent decades.[21] Let us now examine why this is so.

A cornucopia society

In 1957, the sociologist Charles E. Fritz identified and described the convergence behaviour that occurs during the emergency phase of disaster when people and goods flood into the affected area.[22] Commenting on the super-abundant resources that poured into the small Welsh mining village of Aberfan when it was devastated by landslide in 1966, Joan Miller wrote that "we live in a cornucopia society" (Miller 1974). This phrase is so redolent with implications for disaster studies that it demands a critical analysis.

A detached observer of the modern world could be forgiven for thinking that

---

21. Yet the world's chronic problems of food supply and sanitation could be solved for US$13 billion, which is what the US Federal Government alone spends on disaster relief and is a drop in the ocean of international finance.

22. See the work of Fritz (1957), reconsidered by Scanlon (1992).

economics has largely taken the place of religion in late twentieth-century affairs. Economists have become the gurus and soothsayers of material well-being (which despite the dictates of reason has frequently been confused with its spiritual counterpart); they are the prophets of profit and loss. Paradoxically, economics is based on the concept of scarcity (hence the need for economy), but it serves the process of wealth accumulation. Thus, it accentuates the traditional antagonism between richness and poverty, two concepts that are not so relative as to have no absolute meaning: nearly half of the world's monetary resources are controlled by only about 350 rich men and women.

History can be viewed as either the random result of a muddle of conflicting events or as a series of deliberately orchestrated conspiracies to change the world for motives of personal power or gain (in reality it is bound to be a mixture of both). Of the two, it is perhaps easier to see the conspiracy theory in the rise of the global economy. For the time being, capital has won a decisive victory over labour through the twin instruments of maintaining high unemployment and depressing wages by shifting production to where it is cheapest, thus keeping labour forces docile and under threat. At the same time, taxpayers seem to be demanding more and better public amenities for lower payments (which in most cases, besides being an impossible chimera, results in a regressive transfer from income taxation to taxation of goods and services). The two net results of this are a shift from public to private wealth and increasing discrepancies between the rich, or relatively rich, and the poor. World wide, policies that were ostensibly created to make conditions fairer – thus runs the conspiracy – have accentuated both absolute and relative poverty. Simultaneously, the control of primary production and commodity prices by richer countries, multinational corporations and powerful elites has accentuated the North/South divide and created vast areas of deprivation where once it was hoped that they would soon be eliminated.

Realistically, one may ask to what extent the public wealth that is now so enthusiastically being privatized was ever truly public. The answer lies in the extent of democratic participation, the level of official corruption, and the transparency or opacity of decision-making processes, which all vary from country to country and from time to time. In any event, all seem to have been afflicted by the phenomenon that is euphemistically known as "downsizing": the tendency to shrink public expenditures, reduce taxes and provide fewer services and amenities. Much of this is achieved by majority politics – the playing-off of an often small advantaged majority of electors or supporters against a relative minority of disadvantaged citizens.

The nature and effects of disaster aid

Two aspects of this situation affect disaster: the nature of government-sponsored aid and the form of public, private and institutional donations. Considerably less than 1 per cent of the GNP of richer countries is donated as aid and relief to the poorer developing nations, and, despite invocations to spend more, for example in the resolutions of the UN conference on government and development at Rio de Janeiro in 1992, the figures have remained static or have fallen (Fig. 4.4; Middleton et al. 1993, UNDP 1996). The more avaricious governments have claimed that their aid programmes are now more discriminating and better targeted, although both recipient nations and non-governmental organizations have tended to dispute such assertions (Borton 1993). Although only small proportions of international aid are devoted to disaster relief, there are some slight grounds for justifying the developed world's reluctance to spend.

In the words of Eric E. Alley, far too many of the world's resources "have been poured down the bottomless drain of disaster relief, mainly because, when it comes to the expected effects of disaster, myth still overwhelms reality" (Masellis & Gunn 1992). This clearly implies, first, that not all aid money is useful and, secondly, that there are common misassumptions about the nature and usefulness of humanitarian aid. One debatable assumption is that aid is given on the basis of need. There is some truth in this with respect to disasters, in that the industrialized nations will respond to relief appeals made by developing countries, especially if these are specific and not excessively repetitive (in which case "donor fatigue" may set in). But in terms of overall aid programs there is every reason to look sceptically at the criteria by which aid is meted out, if such criteria can be fathomed (Macrae 1998). One need only look to US support during the 1980s for particular factions (government or rebel, according to political leanings) in Central America, or its continued high level of support for Israel, or alternatively US and European support for Iraq in the 1980s, which led to such a disastrous volte face in the 1990s, or to Soviet support for African dictatorships during the Cold War.

Clearly, not only military but also humanitarian aid is a function of geopolitics and influence-mongering (Albala-Bertrand 1993). Some of the most blatantly political combinations of military and humanitarian aid – in Guatemala, Afghanistan, Angola, Mozambique and Nicaragua – have left these countries socially and culturally polarized, in some cases militarily unstable, and economically weakened to the extent that their resistance to natural disaster has declined or collapsed. For example, more than two decades after the December 1972 earthquake (magnitude 5.6) that severely damaged Managua (Nicaragua) the scars were still clearly apparent. Social polarization had been permanently intensified by the earthquake, which brought destitution to the city's poor, but the elite and upper commercial classes soon recovered from it (Ebert 1981).

84

Although disaster aid seems to involve fewer moral and political dilemmas than military and general humanitarian assistance, it can still be withheld in order to bring retribution upon the citizens of uncompliant nations. The debate about sanctions and embargoes is probably endless. In some instances, perhaps in South Africa under apartheid, they may have contributed to positive political changes, but it is highly unlikely that they will ever contribute directly to alleviating poverty and vulnerability. At the time of writing (2000), several years of UN-brokered sanctions in Iraq have caused widespread malnutrition and child mortality without provoking the fall of President Saddam Hussein's regime. Iraq is a seismic country that also suffers from floods and droughts (Fahmi & Alabbasi 1989). But like other nations with regimes considered by the West to be recalcitrant – Iran, Libya and Syria, for example – aid is withheld for reasons that may be highly justified in political and strategic terms but which do nothing to ease the plight and the risks of the ordinary citizen. Yet, humanitarian (including disaster) aid is not strictly tied to political morals; the West has not refused it to, for example, the Congo or Pakistan, where abuses of power are frequent and widespread.

One of the murkier aspects of international aid is the relationship between its military and humanitarian sides. In the worst cases, generous humanitarian aid can be used as an inducement to the recipient country to purchase riotously expensive military technology and hardware. This has been extensively documented in, for example, Britain's relationship with Indonesia and the USA's with Israel. In the former case, the liberal supply of money to build the Pergau Dam also created potential new sources of environmental damage and hazard vulnerability. Although not linked to military sales, a similar situation was narrowly averted in the World Bank's 1980s plan to finance the original Bangladesh Flood Action Plan (Boyce 1990), which would have cost up to US$10 billion and would have kept French engineers busy for years, but would almost certainly have worsened the flood vulnerability, environmental stability and food security situations in Bangladesh. Only a concerted outcry by the best-informed national and international experts succeeded in modifying the plan and making it more benign (Westcoat et al. 1992).

Pure disaster aid – that which is given simply in direct response to relief appeals after the event – is sufficiently limited in size and divested of strategic connotations to be relatively free of constraints on its allocation. However, as public funds continue to shrink in the West, so the amount of such money allocated for disbursement by, for example, the US Agency for International Development and the British Overseas Development Agency has shrunk.[23]

---

23. The overall flux of disaster aid reached a peak over the period 1992–5 following the end of the Cold War and the sudden liberation of governmental funds from military uses (the so-called "peace dividend"). It has since declined (17 per cent by 1997), and this trend looks likely to continue (IFRCRCS 1998: ch. 6).

The temporal problem here is that such agencies are forced by scarcity of funds to be discriminating in their donations, but one never knows what will occur next, and hence how great the needs will be in the next disaster. Disasters, moreover, tend to be seasonal (monsoon floods, hurricane seasons, and so on) or otherwise repetitive (earthquakes and volcanic eruptions): the time series of their frequencies shows distinct trends (usually stable or slightly rising) but with a wide distribution of distinct peaks representing the clusters (IFRCRCS 1994, 1996, 1997, 1998). Hence, there is a strong chance that large outlays on relief will be followed rapidly by further large demands for aid from some other disaster area. It is to be hoped that private donation and the strength of non-governmental organizations will take up the slack left by governmental agencies as they "downsize". Yet there is no guarantee of this.

Disaster aid as boon or bain

Two further aspects of disaster-aid bear examination: its nature and its relationship with development. Relief tends to fall into three categories: goods, services (i.e. manpower) and cash. A major declaration of disaster by one of the low- or middle-GNP countries may engender a response in one, two or all three categories from between 30 and 70 other nations. The response may consist of government-sponsored aid, the involvement of non-governmental organizations (charities, rescue brigades, development groups, missions, etc.), public subscriptions and donations, and corporate shipments. Among members of the public there has been a certain suspicion of direct cash donations to foreign governments and peoples for fear that the money will be expropriated or misused. However, as a general principle, cash is more useful than many forms of goods; it is rapidly convertible and transferable and can be used to buy what is needed locally, thus stimulating the economy of the affected region.

The history of donations of goods over the past few decades is replete with moral outrages. Shipments of clothes have included large volumes of outsize, filthy or ridiculous garments. Imported foodstuffs have been inappropriate in terms of nutritional value, palatability or conservation potential. The best known example of this is the use of skimmed milk (residue of the developed world's cheese production) as an infant feed, despite its obvious lack of the necessary protein, fat and vitamin contents. Lastly, there has been a persistent tendency to send outdated and inappropriate medicines to disaster areas. For example, some drugs supplied to Guatemala after the earthquake there in 1976 were apparently stamped "not to be used after August 1934". After the December 1988 earthquake in Armenia, which killed some 25 000 people, foreign donors sent 5000 tonnes of drugs and consumable medical supplies, constituting 25–30 per cent by value of the total aid supply. But fewer than one third of the drugs were immediately usable: of the rest, 11 per cent were inappropriate medicines, 8 per cent had expired, and much of the remainder were inadequately labelled (with some 238 brand names in 21 different languages). Moreover, little attempt had been made to supply what is really needed to treat earthquake victims (Autier et al. 1990).

In order to keep the record straight, it should be borne in mind that, in most cases, vigorous attempts are also made to do what is right. For example, in the Armenian earthquake aftermath, Dr Eric Noji and his colleagues from the US Federal Centre for Disease Control were instrumental in bringing kidney dialysis machines rapidly to Armenia, as these are essential for the treatment of crush injuries, which liberate potassium and lipids from ruptured cells into the blood and can fatally clog the patient's kidneys (Andrews & Souma 1989). However, useless and redundant aid is debilitating to a disaster area because it absorbs facilities – warehouses, storage sites, dumps, personnel, trucks, and so on – that should be used for genuine disaster relief. At best one is dealing with a grossly inefficient failure to assess needs, and aid should seek to top up, not supplant, what can be acquired within the disaster area. At worst one may be seeing a heartless and cynical form of dumping surpluses on the victims of disaster, or easy way of avoiding proper disposal at home while appearing to be generous. Hence, the surpluses of the cornucopia society can be noxious and detrimental instead of valuable and sustaining.

The Aberfan landslide disaster of 1966 is a classic case in which abundance led to harm (Miller 1974). A mudflow had killed 116 defenceless young children in a poor coal-mining community. The British nation and the world acknowledged their debt to the bravery and self-effacement of miners by sending huge volumes of money and gifts. An enclosed, tightly knit community was riven by the stress of coping not merely with the loss of almost an entire generation of its scions but also with the aid. Fortunately, internal solidarity was not vanquished by the massive influx of help from without, but the aid certainly left its scars.

Much has been written about the effect of donations in other circumstances in distorting local market conditions (Albala-Bertrand 1993). This follows from the tendency, less persistent now than it once was, to supply goods and foodstuffs indiscriminately, rather than in relation to actual shortfalls and what can be generated locally. However, the contrary can exist within disaster areas, in that shortages can occur as a result not of the destruction caused by the catastrophe but of hoarding by producers or wholesalers who seek to drive up prices, especially of basic foods and materials needed for reconstruction. Surpluses, on the other hand, will drive down prices and can cause severe hardship to local producers at a time when their work may have been disrupted by the disaster. Donated goods may be sold, either officially or on the black market. But in all cases, temporary distortions of local markets are unlikely to improve the post-disaster situation. Supplanting market transactions by donation can lead to a very detrimental form of aid dependence. In short, the only viable strategy to relieve hardship among the survivors of disaster is to induce more equity by correcting distortions and gross inequalities in all that is left of the normal mechanisms of supply and demand, production, distribution and consumption. Pure welfare is not the palliative to loss in a disaster (Dudasik 1982, Guarnizo 1993).

Disaster aid and the development process

In one sense disasters represent the failure of development processes, including the inadequacy of policies, procedures, plans and technologies. Thus, it is hardly surprising that the key to the problem has long been regarded as the need to link aid and relief used to mitigate or succour them with development policies. It cannot be done indiscriminately. The risk or occurrence of disaster represents a good opportunity to reassess development strategies in order to build mitigation into them (Cuny 1983, Hagman 1985, Varley 1993).

It is perhaps an oversimplification, and risks being a platitude, that large development projects create risk and vulnerability to disaster, whereas small schemes can more easily protect communities against hazards. But there is an element of truth in this hypothesis. Major economic development has everywhere become a self-perpetuating leviathan. It breeds the technology, the physical plant, the personnel, the corporations, the political networks, the bribes and corruption, even the justifications, that enable it to sustain itself. A survey of the world's major dam disasters tends to bear this out. Dams are useful sources of water and electricity where both are sorely needed, but size appears to be no guarantee of utility or security (Leonards 1987). With the hindsight imparted their failure, the Johnstown (Pennsylvania) Dam collapse (in 1889: 2209 deaths), the Malpassat collapse in France (in 1959: 421 deaths) and the Vajont Dam landslide in Italy (in 1963: at least 1925 deaths) – all can be seen as mistakes in development, which led to death and destruction in natural hazard events. Could the same eventually be true of China's Three Gorges Dam or some of the other major reservoir barriers under construction around the world? In the 1970s, seismic hazards were narrowly averted at great cost at the Auburn Dam near Sacramento, California (Carter 1977, Duffield 1980), and the near failure of the Lower San Fernando Dam in the Californian earthquake of 1971 led to exceptional safety measures in the construction of the Los Angeles Dam that replaced it. In both cases, death tolls in the thousands were averted by timely action, but also in both cases it took an earthquake to stimulate the taking of precautions. In synthesis, sustainable development means not merely protecting the environment but also mitigating hazards by reducing the degree of potential interaction between hazard agents and large outlays of technology, such as major dams.

The leviathan of development is hard to stop and equally difficult to combine adequately with hazard mitigation. Many of the 1987 dead in the 1995 Sakhalin earthquake in Russia, for example, were extracted from the rubble of 80 apartments in only 17 buildings that collapsed completely. Pioneering development of the Russian Far East had not been combined with earthquake mitigation (Alexander 1996, Porfiriev 1996).

Large-scale development – "growth" or "progress" as it is so often euphemistically called – perpetuates itself because it has developed the infrastructure and political and economic networks to be self-sustaining. Once the plant and the capital are there, the land must be developed. If this process cannot be abated, then the best that can be done is to ensure that structural mitigation of hazards is part of the construction process and that planning curbs the worst excesses and channels the development to where it is least at risk and does least harm to the natural environment. The alternative is small-scale growth, aimed at the community level with democratic participation (Guarnizo 1993, Varley 1993). Both initiatives and costs tend to be modest, but there is much scope for building hazard mitigation into the project, especially if the planning and decision making are both transparent and democratic. Such development perpetuates itself by demonstration and example. In general, it cannot sustain large-scale structural mitigation work, so there is a complementary role for carefully directed large-scale development and the two processes are thus not entirely antagonistic.

In summary, the cornucopia society can be considered in terms of both relief and development. The former is a matter of what aid is supplied and how it is administered; the latter is a question of how – and whether – to create and sustain particular patterns of economic growth. A small portion of the world's excess wealth is channelled directly into the relief of suffering after disaster, with mixed results. Some lessons have been learned, often the hard way, by a series of failures, and some errors are constantly repeated. Time and time again, misplaced donation has been shown to worsen matters, however genuine the motivation behind it is. Moreover, an increasing portion of the world's surplus wealth is being poured into the relentless increase of human vulnerability in hazardous areas. It is striking how easily the horn of plenty overflows in ways that exacerbate risk, in contrast to the difficulty of creating a similar abundance of civil protection. It is rather like the way that governments often have a free-handed attitude to military expenditure but a much less liberal one to spending on education.[24] Truly, we are a long way from being able to base human society on a firm regard for safety, rather than a fascination with risk taking.

---

24. See COPAT (1981) for a more sinister model of such tendencies. See also Pugh (1998).

Urbanization and disaster

Much of the world's dynamism, many of its most startling contrasts, and a considerable degree of its natural hazard vulnerability are now found in a specifically urban environment. It is thus opportune to consider disasters in relation to modern urban life and urbanization trends. A large proportion of risk, damage and disaster relief is concentrated into cities, and hence the urban context is crucial to any consideration of modern trends in natural catastrophe and its human impact. Accordingly, the next section will briefly examine the problem of urban living and development in the context of disaster.

Population growth has caused the per capita availability of natural resources to decrease. At the same time there has been a polarization of economic opportunities between rural areas, where they have tended to decline, and urban areas, where they have tended to increase, generating higher rates of consumption of goods, services and energy. The industrialized world is three-quarters urban, the developing world about one-third urbanized, although in Latin America, for example, the proportion is much higher. What is particularly striking is the trend, for the world's population is urbanizing much faster than it is growing (WRI 1994). By the year 2005, half of it will live in towns and cities, and by 2025 4 billion people will be urban dwellers. However, the definition of "urban" is neither fixed nor unquestionable, and it has been noted that rates of urban growth are neither unprecedented nor as fast as has been predicted (IFRCRCS 1998: 11). Already in 2000 there are 34 cities with populations of more than 5 million inhabitants (Figs 4.4, 4.5, 4.6, 4.7). Some of these are Third World "primate cities" that dominate the urban systems of their countries. Although the rate of urbanization in developing countries averaged an impressive 4 per cent over the period 1975–90, it is much higher in primate cities such as Dhaka (Bangladesh; average 7 per cent) and Lagos (Nigeria; average 5.7 per cent).

The pattern in Third World cities has often been one of segregation into small neighbourhoods for the rich and the middle class, and crowding into sprawling shanty towns or tenements for the poor (Fig 4.5). Life can be precarious in the heavily polluted and often insecure environments of the primate cities (Anderson 1992). In contrast, the pattern of urban development has been different and more sedate in western Europe, which has one of the oldest and most mature traditions of urban living. By 1970, 88 per cent of the western European subcontinent's population lived in megalopolitan areas and along international urban corridors and axes, the polycentric agglomerations of well connected towns and cities. There has been a flux of centralization and decentralization. Centuries of urbanization were followed by decades of post-war suburbanization. This in turn was latterly followed by "de-suburbanization" into smaller centres from the large cities with more than half a million inhabitants, where

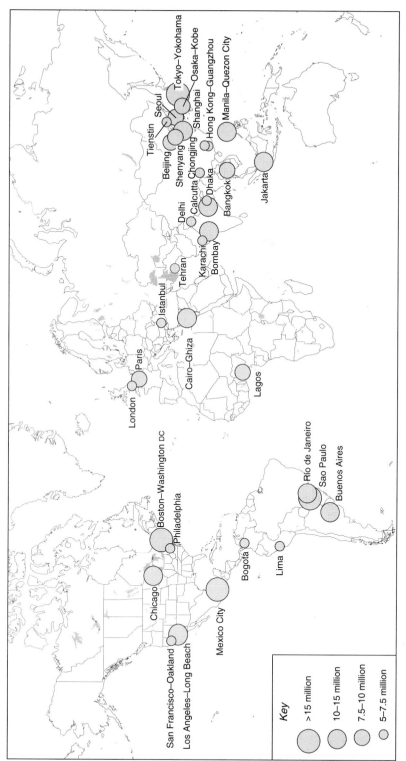

Figure 4.4 The world's largest metropoli.

Key

>15 million

10–15 million

7.5–10 million

5–7.5 million

San Francisco–Oakland
Los Angeles–Long Beach
Mexico City
Chicago
Boston–Washington DC
Philadelphia
Bogota
Lima
Rio de Janeiro
Sao Paulo
Buenos Aires

London
Paris
Cairo–Ghiza
Lagos
Istanbul
Tehran
Karachi
Bombay
Delhi
Calcutta
Chongjing
Dhaka
Bangkok
Jakarta
Tienstin
Beijing
Shenyang
Seoul
Tokyo–Yokohama
Osaka–Kobe
Shanghai
Hong Kong–Guangzhou
Manila–Quezon City

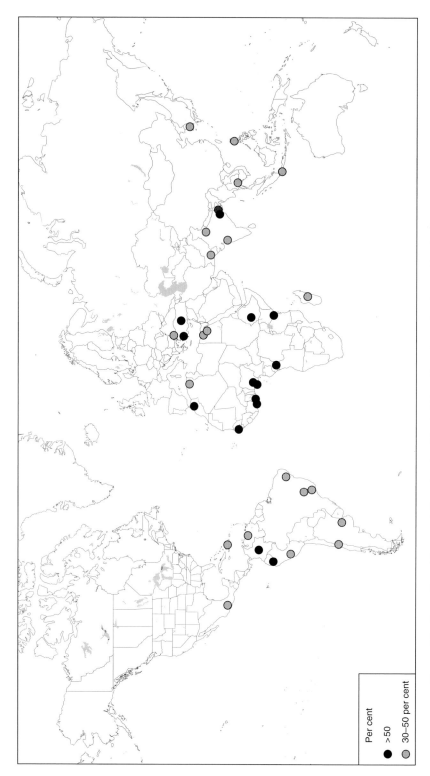

Figure 4.5 Proportion of the population of major cities in bidonvilles, slums, favelas and barrios (source of data: WRI 1994).

Per cent
● >50
● 30–50 per cent

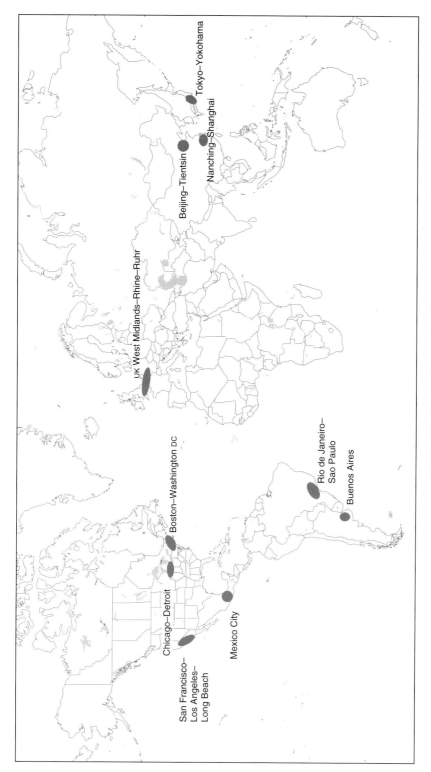

Figure 4.6 The world's megalopoli.

differential access to services had begun to cause acute social polarization. The process has been termed "counter-urbanization" to denote a shift in the relative importance of centres within the overall urban system, not a return to rural life, which is the prerogative of only a tiny minority of professional and leisured people (Clout et al. 1994).

Despite the change in preference towards the medium-size city, where prices are reasonable, services are accessible and the human dimension is not lost, a strong legacy of urban sprawl is associated with the years of suburbanization. In many respects it is being accentuated by recent increases in the decentralization of services that depend on mass car ownership and require more space for vehicle access and parking. This had led to a certain convergence between the two urban forms of the industrialized countries: the Old World multilayered city, with its centuries of constant adaptation of urban functions, and the New World rationalism based upon the revival and reinterpretation of the Greco-Roman uniform plan, with its rectangular *decumanii* and *cardii*, the rectangular pattern of main streets.

In many, perhaps most, parts of the world, large cities constitute the greatest centre of natural hazard risk (Mitchell 1995a, Steedman 1995).[25] In earthquakes, for example, most casualties and losses result from damage to buildings. The loss of 6300 lives and costs of US$131 billion in the 17 January 1995 Kobe (Hanshin) earthquake in Japan illustrate how catastrophic the consequences of such disasters can be for cities (Okimura et al. 1996). The scenario developed for the next major earthquake in Tokyo is even more apocalyptic (Matsuda 1996). The peri-urban and suburban areas have left a legacy of floodable valleys and unstable side slopes (Alexander 1989a), and the spreading urban-wildland fringe is the scene of increased incidence of wildfire, stimulated by depredation of natural vegetation and by arson (Cortner et al. 1990). Many of the largest urban areas are vulnerable to hazards, earthquake, flood, mudflow, hurricane or tsunami, for instance. Disasters may either have widely distributed impacts, as for example in the flooding of huge areas of Dhaka in Bangladesh in August 1988, or, more commonly, cause intense but localized "zones of harm" (Horlick-Jones 1995b: 333). But even when impacts occur in relatively small areas, they tend to affect relatively large numbers of people, and awareness of disasters (although not necessarily of the preceding hazards) increases rapidly as a result of high levels of interpersonal contact and mass media attention (Kelly 1995).

Given this, it is hardly surprising that experts in disaster mitigation have argued that urbanization, perhaps even *metropolitanization*, is one of the principal factors propelling the worldwide rise in disaster losses (Mitchell 1993).[26]

---

25. In one example, Mitchell (1998) observed that the settlements of Randstad (Holland), are individually too small to be regarded as "megacities", but form a collective megalopolis of huge proportions that is threatened in many places by natural and industrial hazards.

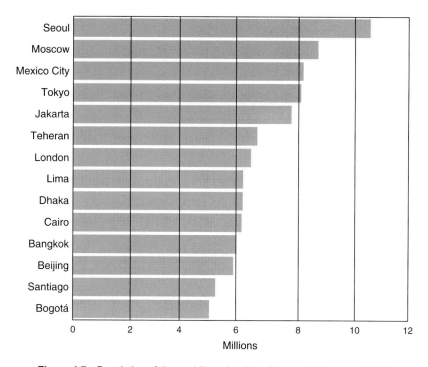

Figure 4.7 Population of the world's major cities (source of data: WRI 1994).

Urbanization has long been identified as a distinctive source of vulnerability to natural hazards, and urban people have long been regarded as less sensitive to natural hazard risks than are rural dwellers, perhaps because they are pre-occupied with more pressing problems of urban life (Burton et al. 1978). The world's largest cities are burdened with the threat of hybrid risks composed of mixtures of natural extremes, technological failures and harmful social activities. In addition, urban conflict often complicates the assessment and relief of disasters, and not merely in the cities of poor, developing countries. At the same time, changes in the pattern of urbanization, for example suburban expansion and inner-city decay, result in shifts in the locus of hazard and risk. Thus, impacts and costs are rising steeply amid highly uncertain social, economic and

26. According to Mitchell (1995a), natural hazard losses sustained by the San Francisco, Los Angeles and Miami megalopoli are equivalent to 1 per cent of US GDP, 1.3 per cent of US national income, 4 per cent of Federal expenditure, one sixth of worldwide corporate profits, or three times the GDP of Bangladesh (all at 1992 levels). These cities have in common a hazardous location, a substantial rate of population growth, complex social relations and a high level of social tension. Moreover, Uitto (1998) noted that half the urban dwellers of the world's 50 largest cities live within 200 km of faults that are known to produce earthquakes of magnitudes greater than 6.9.

technological contexts. The problems are especially acute for major Third World cities, which often suffer a shortage of the most basic scientific information on hazards. Moreover, they cannot benefit from prediction and warning systems unless these are accompanied by changes in standards of living, and access to many different kinds of support system (Mitchell 1995a,b, 1998). In-migrants to cities in developing countries may lose their traditional means of coping with life's hazards, which cannot be adapted from rural to urban living, and finish up in spatially and socio-economically polarized areas, virtual "risk ghettos" (Horlick-Jones 1995b: 330). Hence, in such places as Port au Prince (Haiti) and Luanda (Angola) the *status quo ante* has often seemed as bad as any disaster impact (Kelly 1995).[27] And usually the poor occupy the most hazardous sites – the steep and unstable mountain footslopes, the flash-floodable canyons, and so on.

Yet there are solutions. Local democracy is the key to wise and sustainable development, and the latter is the key to disaster prevention and mitigation (Wisner 1995; IFRCRCS 1998: 16–18). The United Nations and the International Geographical Union have studied the problem of urban hazards and concluded that one of the most promising avenues for mitigation is when hazards overlap with other urban issues, such as poverty, provision of services and renewal of infrastructure (Mitchell 1995b). In this respect, the degree of safety in modern urban environments is very much a function of the politics of risk tolerability (Horlick-Jones 1995b: 330).

But a more critical analysis is warranted. To begin with, what *is* urban?[28] Europe is deeply divided over this question: in Norway and Sweden the threshold is set at a mere 200 inhabitants, whereas in Greece and Spain there are agri-villages with 20 000 inhabitants (Clout et al. 1994: 131–2). In reality, the definition is based not merely on population size and spatial extent but also on the type of administrative unit and the presence of non-agricultural functions. At a more ample scale the concept of megalopolis is open to question, for even though rural space is fragmenting there, it is by no means insignificant, even if it has begun to assume the character of *rus in urbe* rather than *urbs in rure*. Pertinent to the issue at hand, one may ask whether megalopolis has any real meaning in terms of hazard vulnerability (Mitchell 1995a). The answer may be "yes" when the hazard is large, such as a hurricane. The vast interconnectivity and spatial distribution of resources demand concerted action over a very wide area, including perhaps the need for mass evacuation. Thus, some 2.5 million coastal

---

27. According to Cannon (1994), the impact of disasters in developing countries often produces only more acute and extreme forms of workaday suffering.
28. According to Horlick-Jones (1995b: 300), "a city may be regarded as a complex nexus of socio-technical systems, all interacting with shared physical and socio-economic environments". Yet mere definition gives no idea of the extraordinary variety of urban forms and phenomena that exist around the world.

residents were evacuated from Florida and Louisiana in August 1992 before Hurricane Andrew (Pielke 1995). On the other hand, the megalopoli are not continuously urban; open space is liberally distributed, and damage and casualties tend to be spatially concentrated, often in dilapidated inner-city neighbourhoods. Moreover, in places such as Rio de Janeiro and Caracas, it is not so much the urban area that suffers but the precarious peri-urban fringe of *favillas* and *barrios* into which the poor are herded.[29] In a different context the middle-class space-extensive suburban developments of Cincinnati, Pittsburgh and Houston are chronically susceptible to landslide damage, which costs millions of dollars per year (Alexander 1989a).

To what extent is today's hazardousness the result of locational decisions taken in the distant past? Despite some famous examples to the contrary (Noto Nuovo in eastern Sicily, for example; Tobriner 1982) vulnerability to natural disasters has practically *never* exerted a key influence on the decision of where to site cities.[30] The result is a legacy of risk that tends to increase with the steady accumulation of assets. Sewer and storm drainage systems age and fall short of the capacity required by alteration to the superstructure above them. Green space is filled in and flood risk increases (Rasid 1982). Artificial and natural slopes age and weaken to the point of collapse as they are built over, neglected, defoliated or subject to increased weathering and infiltration. Earthquakes selectively "filter out" the older, weaker and more dilapidated buildings (as well as some of the newer but more spectacularly inadequate ones).[31] Yet it has been argued that older-established urban areas suffer fewer and smaller hazard problems than new areas. This highly debatable thesis is based on a comparison of downtown Los Angeles with urbanized canyon mouths and alluvial fans of the San Gabriel Mountain front (Rantz 1970, Cooke 1984). It is doubtful whether it has much validity even there, as much depends on the type of hazard and what has been done to mitigate it; in inner Los Angeles there are still 40 000 pre-code[32] unreinforced masonry structures that are highly vulnerable to earthquake damage (Alesch & Petak 1986). In short, geographical inertia prevails in cities and often at very high cost.

Although certain kinds of rural lobby can be very powerful (well organized farmers' groups, for instance), the urban constituency is the one to which politicians mainly respond. In earthquakes in Guatemala in 1975, southern Italy in 1980 and Armenia in 1988, the paucity of relief in rural areas was both a

29. See Jones (1973) for examples of the impact of natural hazards on the urban fringe of one large tropical city, Rio de Janeiro. For a different perspective see Gruntfest & Montz (1986) and Guerra (1995).
30. See Cooke (1984) for an example that supports this assertion.
31. See Rendell (1985) for a detailed analysis of this process.
32. That is, dating from before the introduction of the first anti-seismic building codes in 1934.

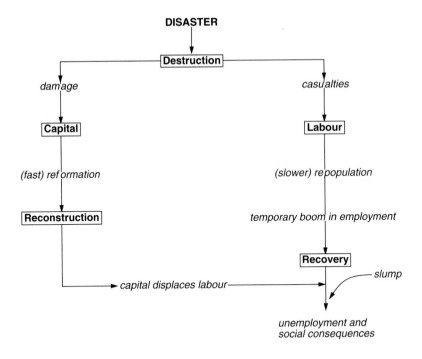

Figure 4.8 Unequal competition between capital and labour in urban areas that experience disasters.

question of official lack of perception and the apparent remoteness of damaged villages (Olson & Olson 1977, De Bruycker et al. 1983, Wyllie & Filson 1989). Thus, urban areas tend to receive more post-disaster aid, and faster, than do rural ones. But this is as far as it goes, for capital and labour are neither very complementary nor very interchangeable in the modern city (Fig. 4.8). One consequence is the tendency of the most powerful institutions, principally financial ones, to rebuild first and in the prime locations (Kates & Pijawka 1977). In times like these, when economic growth is not dependent on full employment, during post-disaster reconstruction labour cannot easily compete with capital, which justifies itself by the specious argument that it is the custodian of the recovery process.

In the industrial world, at least, damaged assets can seldom be rebuilt to the original standards. The reconstruction must take account of new norms and requirements (Sorkin 1983, Key 1995): anti-seismic codes, access for the disabled, fire proofing, modernization and so on. Hence, the impact of disasters and the needs of urban recovery and reconstruction are likely to be very different. This is one reason for the steeply rising costs of disasters and for the exaggerated fears about the impact of future events. A paradox often underlies this: the age of cities tends to belie the fact that short-term planning is carried out at the expense of long-term visions, as a result of political vote-catching expediency (May 1985). This has some unfortunate side effects. For example, it is clear that, by and large, the probability that natural hazards will interact with technological risk is greater in urban than in rural areas. An all-hazards approach to planning is essential (Mitchell 1995a), but it is complicated by the need to consider the risks in groups. Thus, the primary impact might be an earthquake and the secondary one a toxic spill that requires containment and public evacuation in an environment already damaged and disrupted by the tremors. Such problems as this require foresight, ample mitigation funding and sophisticated emergency planning (Whittow 1995).

Examples of urban hazards and their management

At this point let us turn to some examples of hazards and their management in a city of the Old World (Florence, Italy; 400 000 inhabitants) and one of the New World (Palo Alto, California; 104 000 inhabitants; Beatley & Berke 1990). This will highlight the difficulties associated with the problem of limiting hazard impacts in complex modern urban environments and will perhaps suggest some of the solutions. The Italian case involves the added difficulty of managing art treasures and mass tourism; the Californian example refers to experiments conducted with new techniques of data acquisition and hazard mitigation.

On the night of 2–3 November 1966 the River Arno burst its banks in the centre of Florence. Rain had fallen steadily and saturated a wide area of the $8000\,km^2$ basin, and in the urban reaches of the trunk river, which could carry a bankfull discharge of $2500\,m^3/sec$, received $4100\,m^3/sec$. The flood had a calculated recurrence interval of 150 years. The lowest-lying areas of the city, including Piazza Santa Croce, were inundated to a maximum depth of 3.8 m. Thirty-four people were killed; muddy water and heating oil were dispersed over a wide area, many priceless works of art and antique volumes from the National Library were damaged. The economic costs were enormous, although recovery was relatively rapid. Key roles in the emergency operations were played by corps of firemen and groups of volunteers (Alexander 1980).

Since the disaster, progress has been made such that Florence now has some advantages over its 1966 situation, although it is hampered by continuing disadvantages, such as increased traffic congestion. Modern civil protection in Italy stems from a law of 1992 that obliges provinces and large municipalities to mitigate hazards and manage emergencies. Impetus comes from a ministerial department in Rome, led by an under-secretary, who, at the time of writing, is an eminent volcanologist. Florence has acquired modern, technologically advanced emergency operations centre (EOC) with excellent communications facilities and a very detailed municipally based geographical information system. A small but dedicated staff work hard to increase public sensitivity to natural and technological hazards and to promote mitigation. Roles and responsibilities are generally well defined and well apportioned, although there is some similarity between the leadership roles of the provincial prefect and the city's mayor. The EOC can call up very rapidly one of the world's oldest, best established and most extensive volunteer networks. The initiative has the backing of the Department of Civil Protection in Rome, which considers it an important prototype for developments in other Italian cities. At the same time, efforts are under way to involve other municipalities of the Florence metropolitan area (population 750 000) by creating mutual aid agreements and by extending the GIS beyond the city's boundaries. Floods are the principal hazard, but landslides, earthquakes, industrial accidents, terrorist incidents and transportation crashes are also important risks.

However, there are formidable obstacles to hazard mitigation in the Florence area. To begin with, in 1967 a national decision was taken to favour economic growth over environmental protection by suspending planning regulations for one year (the "Ponte" law). Rather than reducing vulnerability to floods, the settlements of the middle Arno floodplain grew by up to 70 per cent during the resulting free-for-all. Structural protection remained stagnant or even diminished, and even by the mid-1990s it could still be judged highly inadequate, despite the construction of a major dam, the Bilancino project, in the mountains north of the city. At the same time, there has been little attempt systematically to assess the vulnerability of the area. The Arno River Basin Authority is doing so for floods and the Provincial Environmental Plan has managed other natural hazards, but an all-hazards microzonation is lacking.

Overall, there is a lack of political (and often public) interest in hazard mitigation, which usually appears to be a less pressing problem than pollution, unemployment, housing shortages, congestion, terrorism, accidents, crime, drugs, immigration and vandalism. Thus, there is a lack of funds and staff for civil protection, even though it interacts strongly with many of the apparently more salient problems. The shortage of resources and lack of impetus means that the city's disaster management strategy is being extended and integrated with that of other jurisdictions only with great effort and in the face of inertia and objections.[33] Within the municipality, disaster planning is evolving, but, at the time of writing, the city's disaster plan is incomplete and is not integrated with those of the local airport or the main regional hospitals. A public evacuation

---

33. Lack of interest and investment in information technology is a particular problem at the municipal level in much of the rest of Tuscany Region outside Florence.

map was published in the local telephone directory in 1989 but was later withdrawn as the generally congested state of the city's roads, their high vulnerability to flooding and the utter lack of public sensitivity to the problem made it patently unusable. The current attitude to managing the public during a disaster is akin to the outmoded philosophy of "keep people in the dark in order to discourage panic behaviour". In point of fact, the risk of disaster is not considered to be very important in Florence and there is a certain official reluctance to assume responsibility for it, although civil protection does have some valiant protagonists in the city. In summary, disasters are not considered very important in Florence and there is a certain official reluctance to assume responsibility for them, although civil protection does have some valiant protagonists in the city.

The problems of Florence, and the painfully slow rate of progress in civil protection, are probably not untypical of what can be found in many European cities. High-consequence low-probability events have little political attraction and little constituency among the accumulators of wealth. All the same, progress is being made and one hopes its effects will be cumulative and will spread around the continent as laggard administrations are compelled to follow the leaders and innovators. There are some signs that the International Decade for National Disaster Reduction has had a catalytic effect in stimulating both research and application in Europe's hazardous urban areas. One hopes for more.[34]

The city of Palo Alto (California) is situated on the coastal plain of San Francisco Bay, 50 km from San Francisco and about 15 km from the north–south line of the San Andreas fault. It includes the campus of Stanford University. Both Palo Alto and Stanford suffered considerable damage during the 1906 San Francisco earthquake (magnitude 8.2). Stores in downtown Palo Alto and the library and chapel at the university were ruined, largely because they were unreinforced masonry structures with little resistance to seismic shearing (Beatley & Berke 1990).

In view of the earthquake risk, the city administration of Palo Alto spent three years in the mid-1990s collaborating with a group of researchers at Stanford and with the consulting firm G&E Engineering Systems Inc., on a study of local earthquake hazards. The principal aim was to use advanced risk assessment and damage simulation to predict the socio-economic impact of a magnitude 8.0 earthquake on the nearest section of the San Andreas fault. The model used a detailed GIS that included information on all 18 738 properties in the municipality. The GIS and seismic response models revealed a sample of buildings that would most benefit from seismic upgrading and which would be most severely damaged without it. As the task of upgrading all buildings in need would be

34. For example, see *Stop disasters* (1996).

prohibitively expensive, cost–benefit analysis was performed on the sample to determine the best way to spend limited funds in order to reduce casualties and economic losses (Eidinger 1996b).

At the time of the study, Palo Alto contained 17 682 wood-frame buildings with a total value of US$2311 million, and 1056 other structures valued at a total of US$105 million, according to tax assessments. The earthquake simulation suggested that moderate damage would occur to 6300 wood-frame residential buildings and 350 other structures. Losses of US$309 million would be sustained by the wood-frame buildings and US$47 million by unreinforced masonry structures (URM structures). However, only one building was expected to suffer a loss of more than US$3.5 million.

Day-time building occupancy in Palo Alto was estimated at 79 600, with nearly half of the occupants in wood-frame structures. Night-time occupancy rises to 104 000, and most people are to be found in wood-frame residential buildings. Casualties estimated for the magnitude 8 earthquake consist of 115 people instantly killed, 160 critically injured (some of them fatally), 250 hospitalized but without threat to life, and 700 injured but not hospitalized. Of the dead, 75 lose their lives in unreinforced masonry buildings and 40 in other structures, but only 15 buildings will be the scene of casualties. These estimates take account of the fact that 10 people died in URM structures to every 3 in wood-frame structures in the 1989 Loma Prieta earthquake (Durkin et al. 1994), and the ratio was 15 to 1 in the Northridge tremors of 1994 (Durkin 1995). The number of casualties was estimated to vary not only with the rate of damage to buildings but also with the quality of search-and-rescue services after the disaster. Seventy-two hours after the event the number of dead might vary between 150 and 190, according to whether lives could be saved by prompt action.

It was predicted that seven fires would be caused in Palo Alto by the magnitude 8 earthquake. Both aqueducts leading to the city would rupture and 300–500 localized breaks in water distribution pipes would result in the rapid drainage of municipal water tanks. If the city honoured plans to construct a new reservoir, this would save 17 houses, but 18 would burn down regardless, according to the scenario (Eidinger 1996b).

The Palo Alto earthquake-loss study has provided many detailed data that are ready to be integrated with the national all-hazards appraisal methodology, known as HAZUS ("Hazards in the United States"; RMS 1993). However, much depends on the validity of assumptions used in modelling, which lack some critical details and which are somewhat questionable, although they may in the end prove robust enough to be an adequate guide to what will happen to Palo Alto when a major earthquake finally strikes. Having designated buildings in the city as critical facilities on the basis of predicted casualties, loss of functionality and high monetary loss, real-time damage assessment is planned. Sensors and data transmitters will be placed in the buildings, and data on earthquake performance will be collected during the next event, and will be used to refine the parameters of the loss-estimation models.

The two cities discussed here illustrate the increasing reliance of urban hazard managers on technological systems in order to reduce risk. This is a general characteristic of municipal and regional hazard management on both sides of the Atlantic and it implies high expectations for investment in the hardware (and the software) required to do the job. Rightly in my view, information management has become the key to mitigation: the city is viewed as a complex amalgam of data to be collected, arranged and analyzed before risks can be reduced or emergencies managed. The many kinds of vulnerability in the urban environment can indeed only be understood with sophisticated information management systems. But it is to be hoped that the human element does not get lost in the race to acquire technology: most inhabitants of cities are not attuned to computerized mitigation and this may mean that they are not so amenable to being managed by information technology. Time will tell, as the new systems are gradually exercised.

Although urbanization has a 9000-year history, it has certainly assumed dimensions that render it a newly transformed force in the consideration of natural hazards. Metropolitan growth is part and parcel of modernity. But it is now time to consider modernism in another light, to examine our current preoccupations in the context of their historical antecedents. In fact, nothing could be more up-to-date than the study of history, which provides the depth, and a good measure of the causal explanation, for many phenomena and events associated with natural disaster.

CHAPTER 5

Past, present and future

An historical approach to modern disasters

The aphorism "those who ignore history are condemned to repeat it" contains enough of a grain of truth to be repeated often. Yet it still smacks of cliché, for, strictly speaking, history does not repeat itself, although failure to heed its lessons can mean that the same mistakes are made repeatedly. One of the great weaknesses of disaster studies has been their lack of a historical perspective. Where history has been tackled, the tendency has been to describe events rather than historical processes. Thus, there is ample scope to investigate the relationship between natural disaster – with its many connotations of risk, hazard, and vulnerability – and the currents of human history. The task is too vast to attempt in a work such as this, although some effort can be made to point the way. But the eventual reward will be a deeper understanding of present and future disasters, and a wider perspective on the evolution of human attitudes to natural catastrophe (Barkun 1977).[1]

No doubt every historian recognizes the fundamental dangers of an historiographical approach to events. To begin with, history is neither a linear nor a cyclical process, although both models have been used to render it more intelligible. Secondly, whether the fundamental forces that drive events are the product of conspiracy or coincidence is a matter of the deepest conjecture; probably there are elements of both. Thirdly, when interpreting history, one risks what Gordon Herries Davies has termed "historical whigishness": the tendency to view past processes with hindsight as a stumbling progress towards the current state of truth and light (Davies 1989). In reality, even the present state of advanced knowledge is a mere marker on an uncertain path that stretches far into the future. Finally, the study of history is replete with predilections: cultural and ethnic biases, the preoccupations of the moment, and the unwitting expression of ingrained prejudices. To some extent it is unavoidable that we view the

---

1. See Short & Rosa (1998) for a different perspective.

past in terms of what concerns us about the present. It is even a legitimate approach if the aim is to cast light on the origins of present-day situations. But it has profound implications for objectivity: the nature of what is considered an orthodox or legitimate interpretation of the past is continually changing in subtle ways. Even history can be obsolescent in this shifting world.

The first task of the historically minded "disasterologist" should be to ask some fundamental questions that will help define the relevance of history to current processes and begin to reveal its lessons (Alexander 1993: 593–602). Here are few of them:

- What impact has natural disaster had upon the course of civilization?
- What is the relationship of natural disaster to the formative processes of society in the socio-economic, political and military spheres?
- To what extent have past disasters acted as vehicles of change?
- How was disaster interpreted in the past and what does that tell us about society's relationship to extreme geophysical events?
- What is the message of past disasters for future trends and future efforts to tackle the problem of natural catastrophe?

As space is limited in the present work, I will discuss only a few general reflections and a couple of examples.

To begin with, the fact that both the Colosseum in Rome and University of California stadium at Berkeley are built on seismically active faults suggests that the historical process is not one of lessons learned. Broadly speaking, history has bequeathed the following to the student of disasters: a long legacy of periodic loss and reconstruction, punctuated by relatively few good examples of innovative mitigation; a scientific understanding of geophysical phenomena that is extremely recent; a much longer intuitive understanding, although one clouded by frequent misconceptions; and a very patchy record of adaptation to natural hazards – a process held in check by lack of population, resources and technology. We will now address in turn the physical and the human aspects of natural catastrophe.

Knowledge of physical hazards in history

The history of the physical side of disaster illustrates the difficulty of gaining any true understanding of extreme natural phenomena without a basis of observational science and its theoretical underpinning, Newtonian mechanics. The operation of gravity, so fundamental to so many processes, cannot be understood intuitively, for it must be known experimentally and by deduction based upon observation. Hence, in 1691 Giovanni Domenico Guglielmini

(1669–1710), who was Professor of Medicine and Natural Philosophy at the University of Bologna, could not adequately explain why rivers meander; although he had many quantitative observations on their flow processes, his understanding of gravity was incomplete (Chorley et al. 1964). Many geophysical processes could not be understood without adequate knowledge of the rock cycle, which depends critically on the role of volcanism. Although volcanic processes and landscapes have long fascinated observers of the natural world – Dante, for example (Alexander 1986c) – since time immemorial, the true role of volcanism in contributing material to the Earth's crust was not appreciated until the advent of romantic and scientific journeying, the age of the Grand Tour (Tinkler 1985).

Thus, in 1751 Jean-Etienne Guettard (d. 1785) climbed the 1465 m peak of Puy de Dôme and saw – for the first time since the Romans – 50 extinct volcanoes (Guettard 1752). Through dogged fieldwork, his successor, Nicholas Desmarest, demonstrated that the columnar basalts of the Auvergne had been intruded into the surrounding granites (Desmarest 1771), and in 1763 Rudolph Raspe discovered that the central European basalts were also volcanic in origin. These observations gave birth to the school of the Vulcanists, who were fewer and intellectually less powerful than their great rivals, the Neptunists, whose view of the origin of rocks was diluvial. But even among the latter there were rumblings about the power of volcanoes. Leopold von Buch (1774–1853) was one German Neptunist who provided a link with the Plutonists, who were next to rise to prominence and who recognized the role of intrusive igneous activity in creating the Earth's crust (Von Buch 1824). Upon visiting the volcanic fields of Naples in 1798 and 1805–6, he was profoundly impressed by the ability of volcanism to create landforms, and he began to search for similar phenomena elsewhere. This led him to the Auvergne and in 1815 to the volcanic island of Tenerife. Thus, he helped show that volcanic rocks were much more widespread than Neptunism allowed. He ended by proposing that orogeny (mountain building) was the result of the intrusion of igneous rocks and their accompanying steam and vapours: basalt jacked up volcanoes, whereas augite porphyry heaved up, contorted and split open mountain ranges by being intruded into the crust, causing the transformation of limestone to dolomite. It was a full chemico-physical theory of the creation of physiographic relief.

It was left to Sir William Hamilton, an enthusiastic amateur volcanologist, to fill in the details as he observed the spectacles provided by the frequent eruptions of Vesuvius in the late 1700s and early 1800s (Hamilton 1772, Knight 1990). The discovery of the geological significance of volcanism was a true revelation: after centuries of conjecture about the Earth's interior, a basis of fact could be built from which to launch more sustainable theories. The process continues vigorously to this day.

The notches on the Nilometer[2] recorded flood levels for nearly 2000 years and the records of Chinese and Russian bureaucrats give a similar perspective on floods, droughts and other catastrophes over vast tracts of the northern hemisphere (Jones 1987). But despite this – the effect of tax, title and yield – scientific measurement is almost entirely a modern phenomenon. In most places it would have been considered pointless or even heretical in past ages. When we look at the precisely observed drawings that Leonardo da Vinci made of floods and water flows, it is difficult to appreciate that he was a completely isolated figure in his time: the impact of his science has largely been felt since its rediscovery in 1850 (Alexander 1982).

In subsequent natural philosophy, the work of the Ancients, for example Aristotle's *Meteorologica*, was much appreciated for its clarity and freedom from dogma, even though it was based on only the barest minimum of observations of the natural world (Aristotle 1952 edition). But it gradually came to be eclipsed in Western society by biblical teaching, which was for centuries a fundamental force for the maintenance of social order. The story of the battle to extend the age of the Earth beyond a nominal 6000 years and to rescind the literal truth of the Book of Genesis has been well told elsewhere (Gillespie 1951, Albritton 1980). The Great Flood, or Noachian Deluge, was to all intents and purposes a misinterpretation of signs, as Leonardo da Vinci saw in 1514–16 (Alexander 1982). Gradually the geological evidence of marine transgressions and regressions, the palaeontological evidence of neritic fossils[3] and the formation of the glacial hypothesis by Agassiz vanquished the notion of a single Great Flood (Agassiz 1967); the modern persistence of biblical fundamentalism is perhaps a problem of primitivism and cultural deprivation (Shea 1983). But the process was by no means one of gradual progression through the patient and cumulative assembling of evidence; the matter was partly confounded by the abundance of genuine evidence for the large floods that can be found around the world.

Biblical portents based upon flood, tsunami and earthquake are still widely used as metaphor and example. They are part of an anthropocentric view of natural hazards in which they are treated as accessories to the human condition, or as afflictions destined to admonish, punish or test one's mettle. Although such moralism may impede the objective understanding of physical processes, it need not be harmful. But there is a more serious problem of the predilection for dogma over clear thinking. This persistent trait threatens to make the Kuhnian paradigm of scientific revolutions (Kuhn 1962) something of a self-fulfilling hypothesis.

One advantage of modern hindsight is that Newtonian mechanics can now

---

2. The ancient Egyptians cut notches in stone to measure the twice-annual flood levels of the Nile. The so-called "Nilometer" was the world's first and longest-lasting flood-stage recorder.
3. That is, fossils from the shallow-water nearshore environment.

be seen as the basis and precursor of the Huttonian uniformitarianism (Hutton & Playfair 1970 edition) that has prevailed for the past two centuries (until, that is, the current millennial resurgence of its *alter ego*, catastrophism). In truth, uniformitarianism has ebbed and flowed throughout the course of Western recorded history (Dury 1980). We will now consider one example of the history of the physical side of disasters – earthquakes – which cannot easily be dissociated from catastrophism. An appreciation of the true nature of earth tremors – if that is what we have been lucky enough to acquire – came very late in the evolution of the Earth sciences.[4] In this respect, earthquakes are anomalous, but the long struggle to understand them casts much light on the difficulties faced in past times by natural philosophers and natural scientists, who were often men of formidable intellectual stature (Davies 1989).

To get to grips with the history of seismology, we must first delve a little into the background of contemporary natural philosophy. This rests upon a paradox: catastrophism could exist within apparent uniformitarianism. According to Plato (429–347 BC), who together with Aristotle provided the basis for theories of the Earth for more than two millennia, the passage of time was cyclical, not linear. This concept stemmed from the idea that permanence and perpetuity gave things identity, hence enduring phenomena that changed must be recurrent, and change itself was an inferior order of reality. When compared between different cycles, there was an absolute lack of uniqueness in events. Plato's "great year" (*magnus annus*) lasted 36 000 years, the time that ancient astronomy reckoned necessary for all heavenly bodies to complete their orbits and realign themselves (the cycle proposed by Hipparchus in the second century BC lasted 26 000 years and thus reflected modern astronomical cycles more accurately). Plato's "great year" consisted of a *magnus hiems*, or "great winter" and *magna æstas*, or "great summer". The former was a period of terrestrial floods and the latter of drought and fire (Adams 1938).

These concepts point towards a cyclical equilibrium in which there was a natural order of things. The concept of "element" meant not an indivisible substance, as it does today, but a phenomenon with immutable properties. Only five elements were recognized: in their natural order outwards from the centre of the world they were earth, water, air, fire and æther. Earth was derived from water, air and fire, and æther was nebulous in both substance and conception. Only eight heavenly bodies were acknowledged to exist: the Moon, the Sun, Mercury, Mars, Jupiter, Saturn, Uranus, the orb of the stars, plus the outer

---

4. As it is difficult to measure and model seismogenic processes, which occur under conditions of extraordinarily high pressures and temperatures, seismology does not have all the answers. The sheer invisibility of earthquake-generating mechanisms made them the subject of centuries of speculation and, even in the twentieth century, the measurement of them remains imprecise.

firmament. It was the varied alignment of these that made the elements unruly and hence drove the processes of change at the Earth's surface. Permanence begat change, order gave rise to disorder.

The history of mankind's view of earthquakes is intimately bound up with how it interpreted volcanism. The Greek geographer Strabo (*c.* 63 BC to *c.* AD 20) regarded fire as the prime mover of surficial processes, especially elevation and subsidence of the land, and he considered the volcanoes he saw on his travels – Etna, Stromboli and Vesuvius – to be Earth's "safety valves" (Strabo 1969 edition). Similarly, the 37 volumes of the *Naturalis historia* of Pliny the Elder (AD 23–79) contain many references to earthquakes, floods and landslides, although imbued with a fair degree of animism (Pliny 1956 edition). But was there water or was there fire at the centre of the Earth? Like Strabo, the elder Pliny saw volcanoes as evidence of Earth's internal fires, a view that may have led him to scientific martyrdom when he perished on the beach at Stabiæ in AD 79 during the eruption of Vesuvius, to which he had gone to experience it first hand.

Despite the prevailing philosophy of uniformitarianism, Classical thinkers had no doubts about the catastrophic impact of seismic activity. For example, Plato (in the *Critias*) stated that Atlantis had been engulfed during earthquakes (Plato 1962 edition), and Strabo (in the *Geography*) was somewhat responsible for the idea that earthquakes cause underground caverns that may eventually swallow up cities.

This was the setting in which earthquakes were considered; no other natural phenomenon generated such controversy. In the progress of Western thought, all four of the principal elements were variously invoked:

- Water – Thales (sixth century BC) ascribed the tremors to seiching of the fundament or, in other words, large-scale oscillation of Earth's internal reservoir. Speleology seemed to have enormous promise, for cave systems were judged to be practically infinite in length.
- Fire – Anaxagorus of Clazomenae (500–428 BC) explained earthquakes as the result of underground gas explosions.
- Earth – Anaximenes attributed the shaking to huge cycles of desiccation and rehydration that alternately split and swelled the Earth's crust, and Lucretius, in *De rerum natura*, argued that subterranean landslides were the culprit.
- Air or vapour – Archelaus thought that these were compressed in underground caverns and spontaneously emitted, and Seneca (in *Quæstiones naturales*) believed that the shaking resulted from blasts of wind caused when underground caverns collapsed (Seneca 1971–2 edition). On the other hand, Aristotle believed that the evaporation of underground waters in contact with the eternal fire at the centre of the Earth led to windy "exhalations", which shook the ground with particular violence where the

subsoil was porous. Indeed, such terrestrial flatulence allowed Ovid (in the *Metamorphosis*) to theorize that mountains were the "puffed up blisters on the skin of the Earth".

So inventive was the debate, and so elusive the cause of earthquakes, that by 1571 the scholar Augustus Galesius was able to summarize 28 different theories on what made the ground shake: natural versus supernatural causes, definite (astronomical) versus indefinite (demons, gods or comets), extrinsic versus intrinsic, single or multiple, creation or destruction (Adams 1938).

Authors from Petrarch (*Lettere senili*) to Kant (*Universal natural history*) lamented the dangers and devastations of earthquakes. Whatever their causes, they seemed to demonstrate that the terra firma underfoot was, paradoxically, hollow. This popular misconception led a German Jesuit, Athanasius Kircher (1602–80), to theorize, observe and experiment until he could propound a theory of the world in which fire and water vied with each other in a gigantic and spectacular *Mundus subterraneus* (Kircher 1664–5), which he illustrated with detailed diagrams of the Earth's interior. A great fire at its centre communicated through branching fissures and passages with smaller fires not far below the crust. Water from the oceans percolated down to caverns beneath the mountains, termed *hydrophylacia*, where it was heated by the peripheral fires and distilled, as in an alembic (or retort) of blown glass. The water could break through the surface as hot or cold springs and the fire as volcanic eruptions. Kircher had visited Mount Vesuvius when it was active and had travelled through Calabria in March 1636, when a major earthquake occurred there, and both phenomena had impressed him greatly. Through analogy with gunpowder, then widely used in warfare, he regarded earthquake shocks as a result of the explosion of sulphurous and nitrous gases pent-up in the hydrophylacia and heated by the same fires that distilled the underground waters.

In essence, this explanation was a more sophisticated version of a prototype expounded by Anaxagoras and many luminaries since. As an example of the latter category, in 1761 John Mitchell suggested that deeper fires cause greater tremors; but volcanoes are relatively free from major seismic activity because they vent the subterranean vapours whose explosion induces the ground to shake. Another theory, gleaned by Mitchell from the effects of the 1692 Port Royal (Jamaica) earthquake, was that the nearer the earthquake occurs to the mountains, the greater the shaking that it causes. In other words, the inflation of mountains leads them to be inherently unstable, as Ovid had predicted in the *Metamorphosis* and Kircher in his *Mundus subterraneus*. Moreover, underground fires hollow out caverns as they burn their way through flammable rock material. The burning contorts the strata and enlarges the caverns until they can no longer support their own weight. But before this, percolating groundwater will flash and cause a spontaneous exhalation, with a rapid vibratory motion (the earthquake), which will split the strata along faults and cause the roof of the cavern to collapse. The vapours do not necessarily escape to the atmosphere, as the cooling action of percolating seawater creates a solid rock barrier above the burning cavern, and the best route of escape is often lateral, over great distances along subhorizontal strata (Alexander 1989b).

Such ideas were applied to the Earth's surface by the English physicist Robert Hooke (1635–1703), who was a confirmed catastrophist. He came up with an immensely complicated treatise that seems to attribute most of geology to seismic causes (Oldroyd 1972). For instance, his ninth proposition runs as follows:

> It seems not improbable, that the tops of the highest and most consider-
> able Mountains in the World have been under Water, and that they them-
> selves most probably seem to have been the Effects of some very great
> Earthquake. . . . That it seems not improbable, but that the greatest part
> of the inequality of the Earth's Surface may have proceeded from the
> Subversion and tumbling thereof by some preceding Earthquakes.
> (Hooke 1705: 291)

Hooke's death occurred on the threshold of the century in which a form of scientific catastrophism came to vie with the religious kind and to build a continuous dialectic with the opposing view, gradualism. Indeed, the baroque culture out of which the Enlightenment grew was based upon the tension of opposites. The middle of the century was marked by the earthquake, tsunami and fire that struck Lisbon on All Saints' Day in 1755 and killed more than 60 000 people. The Optimist schools of philosophy, which had argued that by glori-fying God's happy purpose nature provided the best of all worlds, was thrown into disarray. Nature suddenly appeared to have been made as much for the tor-ment of humankind as for its delectation, and the impression was reinforced by the earthquakes that killed 29 500 people in Calabria in 1783 (Alexander 1989b).

Catastrophism seems to have owed its persistence in no small measure to Mediterranean volcanism. Hooke took historic eruptions in Greek and Italian territories as his model for morphogenesis. Even Charles Lyell, who was as much a Grand Tourist as many of his most influential forebears, recounted the Lucretian and Senecan explanations of seismicity based on fire, wind and subterranean cavern collapse (Lyell 1830–33). He drew heavily on Sir William Hamilton's monograph *Campi Phlegraei* (on the Phlegraean Fields of the active caldera at Pozzuoli, west of Naples), in which volcanism commingled with seismicity in careless abandon.

In 1745 the Leyden jar (a means of generating static electricity) was invented, and William Stukeley seems to have been the first to suggest electrical discharge as the cause of earthquakes. The odour of lightning bolts created in laboratory experiments convinced Stukeley that metalliferous particles were present in the underground vapours, which had been generated by the action of electrical energy on various ores and minerals, including phosphorus and arsenic (Stukeley 1750). Electrical pulses could be transmitted through the Earth along mineral veins and would cause sparks underground that ignited subterraneous vapours. As mineral veins were apparently more common in the roots of mountains, the generation of vapours would be greatest there (according to Isnard 1758), and this created the caverns, which caused earthquakes by collapsing when the force of pent-up gases was too much for them to sustain. This idea remained popular for at least 50 years. Even Robert Mallet, who did so much in the mid-nineteenth century to found seismology as an observational science, attributed earthquakes to underground steam explosions. He also believed that the non-elastic component of seismic waves was responsible for orogeny (Guidoboni & Ferrari 1987).

Thus, theories of earthquake genesis that pre-date modern seismology can be classified broadly into vulcanist and electricist, with a variable amount of overlap concerning the details. Vulcanists such as Kircher and Mitchell believed that subterranean fire was the primary motor for earthquakes, whereas electricists such as Stukeley and Pietro Vannucci (1789) preferred underground static discharges. Steam or gases could figure in either hypothesis, as could cavern collapse under the mountains.

Modern seismology

The foundations of modern seismology were laid by the experimentation and data gathering of Robert Mallet, the inventor of the isoseismal, and John Milne (d. 1913), the father of the seismograph. But it was the Japanese geologist Bunjiro Koto who made the first significant step towards solving the problem of the origin of seismicity. After the 1891 Mino–Owari earthquake, surface faulting was so pronounced on Honshu Island that Koto began to re-evaluate the assumption that earthquakes gave rise to ground breakage, and reverse the direction of causality – a revolutionary discovery.

The history of seismology is perhaps unique in that it required the beginnings of a theory of global tectonics before the fundamental processes at work, crustal stresses, could fully be appreciated. In this we see a happy symbiosis between the study of earthquakes in the twentieth century and the rise of modern global geology. Until the pattern of world seismicity has been established, there was no chance of appreciating its significance.[5] Local studies did not really offer the sort of "window upon the inner world" that early students believed

---

5. Both Mallet and Milne had much to do with this.

they were looking through (and that may also be true of a good many modern developments in science, including soil erosion and desertification studies). Thus, plausible theories turned out to be entirely erroneous. Here we see the danger of over-reliance on a single method. Early theories of seismogenesis had to be entirely deductive, as there was no way to observe the Earth's interior. But although deduction is a strong method, it becomes weak when deprived of the inductive observations that catalyze it. Natural philosophers were forced to fill in the details of their theories by dint of imagination, as shown by the entirely fictitious maps of the Earth's interior produced by Athanasius Kircher. This led them to seek confirmation of the themes selectively, or in other words to look for phenomena that fit the bill and discount those that did not. Yet despite this, the origins of seismology show a delicate interplay of theory (by, for example, John Mitchell and Robert Hooke) with practice (in the work of, for instance, Giovanni Vivenzio and Robert Mallet). Paradoxically, by firing the enthusiasm of the fieldworkers, the formulation of seismological theory stimulated the collection of practical observations, although the observations have proved far more enduring than the theories that they were fed into, a situation that will undoubtedly be replicated with many modern scientific endeavours.

Attitudes to disaster in history

We have dwelled at length on the physical aspects of disaster and it is now time to consider the human side. It is as well to remember that, with the exception of its impact upon Bengal and China, for much of human history natural catastrophe has been an unexceptional source of mortality and destitution. It must be seen in the context of pestilence, epizootics,[6] crop disease, warfare, oppression and civil strife (Jones 1987). It is perhaps unrealistic in the light of such phenomena to isolate the role of natural disasters in keeping life expectancy low, mortality high and resources scarce in relation to population levels. Moreover, as I have argued elsewhere (Alexander 1993: 593–602), there is virtually no evidence that natural disaster has caused the end of particular cultures or civilizations. Therein lies another paradox: the individual could be extremely vulnerable to nature's violence, but the social body was not. Hence, the defencelessness of humanity in the face of extreme natural events has been an individual rather than a collective phenomenon: each disaster led to death tolls that a subsequent surge in fertility among the survivors would eventually redress, and there always were survivors.[7] Given the inventiveness of human beings, and the species' extraordinary capacity to reproduce and survive, it is hardly

6. Epidemics that break out in animal populations.

114

surprising that disasters do not put an end to civilizations, although if a nuclear winter or asteroid impact occurred this might no longer be true.

The history of attitudes to disaster is another matter. The failure of attempts to prevent nature from wreaking its violence on communities of people led it to be viewed as implacable, but it did not stop efforts to propitiate it by prayer, sacrifice or any feasible means. The Ancient scholars and their Arab successors adopted a fairly optimistic attitude to nature, but the Western Judaeo-Christian tradition has often preferred to view the world "through a glass darkly". Until the arrival of Romantic tourism in the 1700s, natural scenery was considered with at least suspicion and often loathing, this despite the ministrations in 1290 of Petrarch, the mountain climber and Kenneth Clarke's "first modern man" (Clarke 1953). Hence, the "warts and superficial excrescences" that the English proto-geologist and minister of the church, John Ray, saw in mountains when he wrote in 1653 (Ray 1692).

According to Maravall (1979), the culture of the Baroque grew out of the tension of opposites, a dialectic that was best expressed in the acerbic polemics of Voltaire (Carozzi 1983). This stimulus pushed the current of thought from dogma to Enlightenment. Although, as noted above, Enlightenment rationalism suffered a great setback in the Lisbon earthquake of 1755, it nevertheless had given educated men a taste, indeed a thirst, for acquiring and sharing knowledge. The birth of Rome's Accademia dei Lincei and London's Royal Society ushered in the modern principle of "knowledge for its own sake". Science, with its principles of objectivity and replicable experimentation, was the only means of making sense of the avalanche of facts and measurements that had begun to pour in.

Yet apart from the birth of modern science, there are other reasons why we can discern the antecedents of the modern attitude to disasters in eighteenth-century Europe. While the relationship of peasants and the urban poor to extreme events remained elementally brutal, the leisured classes had begun to develop a new form of coexistence with hazard, in which chance and risk played a much greater and much less deterministic role than before. The age had begun in which risks would be taken, not blindly, but for clearly perceived benefits. The increasing division of labour, accumulation of surpluses and investment of capital began to make this possible. Hence, the following example illustrates a case of what we might term "inspired risk taking", in which the resources of art and artifice were pitted against an unpredictable volcanic hazard.

---

7. Placanica (1985) chronicled the rise of fertility and human sexuality that followed a heavy loss of population (29 500 deaths) in the 1783 Calabrian earthquakes.

PAST, PRESENT AND FUTURE

The historical geography of the Vesuvian villas

Let us now consider how the drama evolved in history with respect to one remarkable example: the Vesuvian villas of southern Italy.

The history of the Vesuvian villas offers a remarkable legacy. To begin with it illustrates the growing attraction of hazardous areas. The example in this case is a rather extreme one, as the hazard itself attracted noble settlers and their retainers, although of course other scenic and climatic benefits accrued. Space is at a premium in the Gulf of Naples area and the pace of urbanization has not slowed since the heyday of the villas, volcanic risks notwithstanding. Secondly, the villas now constitute an architectural heritage worthy of preservation (Ente per le Ville Vesuviane 1981). This means that they must be protected as far as possible from both the continued volcanic risk and any modern extension of the same urban growth pressures that led to their construction in the first place, and this against a background of acute overcrowding, social tensions, and high potential natural hazard risk. This sort of dilemma is not unusual elsewhere in the world: for example, wherever earthquakes threaten groups of historic buildings.

In summary, although most of the building stock in the circum-Vesuvian area dates from the latter half of the twentieth century, no building of quality comparable to that of the eighteenth-century villas has been constructed since the First World War. It is tempting to view this as further evidence of the decline of civilization, at least at the regional level.

But let us start with the hazards and risks. In the seventeenth century the eastern, southern and western flanks of Mount Vesuvius were covered with pine woods, orchards and market gardens. Population densities were much lower than at present and large sections of the land were uninhabited (Museo Pignatelli 1985). In the eighteenth and nineteenth centuries at least 143 villas and noble palazzi were constructed on the lower slopes of the volcano, as the eighteenth-century equivalent of investment properties and, concurrently, holiday homes, occasionally with secondary functions related to agriculture or hunting. For several reasons, this group of properties now constitutes a unique and precious cultural heritage. First, it contains some of the best examples of Neapolitan baroque, rococó and neoclassical architecture, parks and gardens (Pane et al. 1959). Secondly, it is a unique example of urban planning and human settlement, one that represents a conscious desire of the builders to pit their ingenuity and wealth against the hazards of the volcano (De Cunzo 1979). Thirdly, the environment and function of the villas have both been transformed almost totally since the main period of their construction, 1725–85 (Fig. 5.1; De Luca 1974).

At least 40 eruptive phases have been recorded on Mount Vesuvius during historical times (Figs 5.2, 5.3; Phillips 1869, Carta et al. 1981). In volcanological terms, eruptive activity has ranged from the Strombolian (mildly explosive),

Figure 5.1 The modern environment of the Vesuvian villas: Mount Vesuvius from Portici on the Bay of Naples, one of Europe's most densely populated and heavily congested areas.

through Vulcanian and Vesuvian phases, to the production of a Plinian (highly explosive) tephra column. Hawaiian-style activity has produced lava fountains along eruptive fissures, and summit eruptions have been complemented by the creation of parasitic cones in association with the fissures. The main eruption of historic times was undoubtedly the Plinian one of AD 79, which deposited 2 m of tephra as far away as 15 km to the southeast and sent ashflows and mudflows (lahars) to the south. This eruption also blasted away the upper rim of the mountain, then known as Somma, leaving a rampart (Mount Somma) to the north of the present summit. This has acted as a protective barrier for subsequent eruptions, which have consequently not had much effect on the communities to the north of the volcano (Scandone 1977, Barberi et al. 1990).

Major eruptions also occurred in 1631, 1794 and 1906, and caused significant damage to the settlements on the eastern, southern and western flanks of the volcano. Thirteen towns were affected by the first of these events and about 4000 people were killed. The southern town of Portici was destroyed for the second or third time and Torre del Greco, which is situated astride an active fault that runs towards the summit, was largely obliterated by viscous lava flows from subsidiary cones along the fault. Molten lava reached the Bay of Naples and caused the sea to boil and steam. At night, lightning played over the crater of Vesuvius, and clouds of gas and ash were tinged a luminous red against the blackness of the sky. During the next 200 years Torre del Greco was damaged three more times by lava flows (De Seta et al. 1980).

Figure 5.2 Two contemporary views of eighteenth-century eruptions of Mount Vesuvius. Facing page: the 1794 eruption; above: the 1731 lava flows.

In retrospect, the eighteenth and nineteenth centuries were periods of much greater activity than the twentieth century (they involved at least 13 and 19 eruptive phases, respectively, compared with only two since 1900). But against the destructive effects must be balanced the advantages: soils were revitalized and the eruptions provided a compelling and oft-repeated spectacle for the curious (Museo Pignatelli 1985). Ignimbrite, tuff, basalt and tephra from prehistoric eruptions provided the main local building materials: Neapolitan yellow tuff and Neapolitan grey tuff for easily workable building stone, the basalt of lava lakes and fountains – known as piperno – for a hard ornamental and paving stone, and pozzolano, a mixed tephra, for cement (Ippolito 1985).

By the end of the eighteenth century the circum-Vesuvian zone contained a remarkably high density of architectural masterpieces and gracious ornamental parks and gardens (Strazzullo 1968). This was the result of several favourable circumstances acting in concert.

In 1595 about 17 000 people lived on the lower slopes of the volcano. This figure is remarkably high for the time and it reflects the role of Naples as a major European capital supported by a rich agricultural hinterland. The eruption of 1631 not only caused mass casualties but led to temporary abandonment of much local property: pre-eruption population levels were not regained until 1688 (Giustiniani 1969). Thereafter, land use intensified and the population multiplied rapidly: in 1778 the internal slopes had a population of 27 000 and the coastal strip from Portici to Torre Annunziata was home to 37 750 people; 50 years later the figures were 36 300 and 63 270, respectively, and by 1976 they were 167 000 and 355 000 (Table 5.1; Petraccone 1975).

Hence, by 1700 a stable labour force existed in the agricultural and artisan sectors. The first half of the eighteenth century was also a period of remarkably good harvests (De Cunzo 1979: 86) and steadily rising agricultural prices (Macry 1974: 290–93). Neapolitan noblemen who owned large estates in the Capitanata, Terra di Bari or elsewhere were able to accumulate substantial surpluses of capital, which they could then invest in the hinterland of Naples. However, famine and plague recurred in 1759–64, which caused a crisis in agricultural production and a temporary or permanent halt to the construction of many villas (Luca 1969).

The birth of the villas was also intimately linked to the main historical events of the period (Petraccone 1981). From 1707 to 1734 Naples was a vice-regency of Hapsburg Austria. During this period the main villa to be constructed (in 1711 at Portici) was built by an Austrian nobleman, Emanuel Maurice de Loraine, Prince of Elboeuf (De Seta et al. 1980). Independent monarchy was restored in 1734 with the accession to the throne of Charles of Bourbon. Three factors then encouraged the Vesuvian building spree: the regime's policies with

Table 5.1 Census data for the circum-Vesuvian municipalities.

Municipality	Distance to Naples (km)	Orientation*	Altitude (m)	Area (km²)	Urban area (km²)	Urban area (%)	Population† 1960	Population 1976	Population 1990	Population density 1990 Municipality	Population density 1990 Urban area
Boscoreale	24	SE	55	11.20	2.03	18	17215	20895	27319	2439	13458
Boscotrecase	23	SE	92	7.18	1.76	25	21027	20488	11299	1574	6420
Trecase	21	S	90	6.45	1.25	19	–	–	9581	1485	7665
Cércola	10	NW	75	3.74	2.48	66	11071	16249	16901	4519	6815
Massa di Somma	12	NW	77	3.50	0.48	14	–	–	5492	1569	11442
Ercolano	11	SW	44	19.60	6.98	36	45148	55013	60869	3099	8720
Ottaviano	22	NE	220	19.90	4.97	25	16320	19664	22276	1122	4482
Pollena Trócchia	15	NW	148	8.11	1.65	20	5385	7475	12216	1506	7404
Portici	10	W	26	4.52	3.96	88	50373	83135	67824	15005	17127
S. Giorgio a Cremano	9	W	52	4.11	3.46	84	22423	61193	62168	15126	17968
S. Giuseppe Vesuviano	25	NE	85	14.10	5.76	41	20584	23674	25953	1841	4506
S. Sebastiano al Vesuvio	12	NW	173	2.60	1.47	57	3464	7115	9499	3653	6462
Sant'Anastasia	13	NW	149	18.80	3.47	18	16780	20925	26897	1434	7751
Somma Vesuviana	18	N	165	30.70	7.75	25	17887	21523	29215	950	3770
Terzigno	27	E	104	23.50	2.97	13	10160	11337	13574	577	4570
Torre Annunziata	21	SE	10	7.33	5.10	70	54800	57427	50343	6868	9871
Torre del Greco	14	SW	38	30.70	7.05	23	77576	98523	101456	3309	14391
TOTALS	–	–	–	216.00	62.60	–	390213	524645	552882	–	–
MEANS	16.9	–	96	12.70	3.68	38	–	–	–	3887	8990

* Orientation with respect to Vesuvius.
† The population of the area rose 34 per cent from 1960–76 and 42 per cent from 1960–90.

respect to building, the example set by the sovereign and the cultural atmosphere of contemporary Naples (Alisio 1979).

After the restoration, hostility towards the overprivileged Neapolitan clergy began to grow. Finally, on 9 April 1740, Charles's minister Brancone issued an encyclical forbidding all ecclesiastical building programs, which thereafter required a government licence for their completion (De Luca 1974). At the same time, planning provisions and tax concessions were implemented in order to liberalize secular building. Charles had already set the example by beginning the most opulent of the villas, the Reggia di Portici, in 1738, contemporaneously with the Reggia di Capodimonte in Naples, and 13 years before the start of building at the Reggia di Caserta. This set the fashion for the next half a century (De Filippis 1971).

The construction of the villas was also influenced strongly by contemporary cultural developments and their antecedents. Planning and the various architectural movements received a stimulus in the seventeenth century from the work of the Spanish Viceroy, Pedro De Toledo, which was to last a century or more (Strazzullo 1968). The further development – perhaps overdevelopment – of the baroque into the rococó style, with its emphasis on broken pediments, compound curves and richly ornamented stucco, had no greater expression than in eighteenth-century Naples, especially in the 130 Vesuvian villas extant from this period (Blunt 1975). The excavation of Herculaneum, begun in the 1710s by the Prince of Elboeuf, was resumed in earnest in 1738, and that of Pompeii in 1748. Although this can be regarded as a form of "robber archaeology", the quality and refinement of the artefacts brought to light strongly influenced contemporary tastes and the cultural milieu (De Cunzo 1979). Finally, there were many talented architects in Naples in the first half of the eighteenth century, including Domenico Antonio Vaccaro, Ferdinando Sanfelice, Ferdinando Fuga, Luigi and Carlo Vanvitelli, and Mario Goffredo. Although their involvement in the villas was slight (Fuga in La Favorita, L. and C. Vanvitelli in Villa Campolieto, Ercolano), their designs were reinterpreted in many of the villas (De Seta et al. 1980).

Essentially, in true Palladian style, three types of Vesuvian villa were constructed (Pane et al. 1959). Villas to the north of the volcano and high on the southern flanks had a predominantly agricultural function, which their form tended to reflect. Villas clustered along the coastal strip were built either purely for recreational purposes or for recreation and market gardening. Thirdly, elegant townhouses were constructed in areas of denser population, nine of which are extant (Gleijeses 1980).

The second group of villas appears to be the most important (Pane et al. 1959). Many of these were constructed along the Strada Reggia delle Calabrie, the main road that runs from Naples southeastwards along the coastal plain beneath the volcano; indeed, the Royal Palace at Portici was built astride this highway, a mondanité (worldliness) that would have been considered most improper by northern European monarchs. The road acquired the title locally of the Golden Mile, because of the elegance of its buildings. Villas with a predominantly recreational function were built with formal gardens, elevated terraces, loggias, exedre and coffee houses. Many had a double façade, with terraces that had ornamental screens which framed the views of the Bay of Naples and Vesuvius. The inclined vista, from the mountain to the sea, was such a strong factor in the design of the villas that some (such as Villa Bruno-Prota at Torre del Greco) were even constructed so as not to obstruct the axis of their own parks (Fig. 5.3), which was exactly orientated towards the summit and the coast. Various solutions to the enclosure of courtyards (L, double L, C, double C, rectangular, etc.) were then employed in accordance with the main function of each villa (De Seta et al. 1980).

In short, the eighteenth-century fashion for constructing villas on the flanks of Mount Vesuvius was a game of chance, in which accumulated surpluses of capital were staked against the risk of property losses in an eruption, with the added benefits of the climate, scenery, coastal amenities, proximity to the city and spectacle of volcanic phenomena:

Figure 5.3 Villa Bruno-Prota, Ercolano, a Vesuvian villa designed to avoid obstructing the line of sight that runs from the seashore to the top of the volcano.

Often the volcano erupted and, in truth, what nobleman of Europe could offer from his own balcony such a spectacle! The thrill of fear, the little statuette of St Jannarus [patron saint of Naples] with head bowed and hand raised in protection! (De Cunzo 1979: 86)

Construction and modification of the villas continued throughout the nineteenth century and finally ceased shortly before the First World War (Alisio 1984).

In 1750 Giovanni Carafa, Duke of Noja (1711–68), began a detailed map of Naples and its hinterland (*Mappa topografica della città di Napoli e de' suoi contorni*; an extract is shown in Fig. 5.4). The work, which was published posthumously, took 25 years and 738 184 ducats to produce. The 35 folios of the 1775 Noja map and its derivatives constitute a reliable and much-quoted source of information on the villas existing at this time (Almagià 1912, 1913, De Seta 1969). Many changes have since taken place.

A royal decree of 19 June 1836 authorized the construction of Italy's first railway, which was opened from Naples to Portici on 4 October 1839 and from Portici to Nocera on 18 May 1844. This, the subsequent narrow-gauge circum-Vesuvian railway line and the A3 autostrada trunk road of the 1940s have all had a profound impact on the environment of the villas (Alisio 1984). For example, the main railway line, the principal rail link between the North and Sicily,

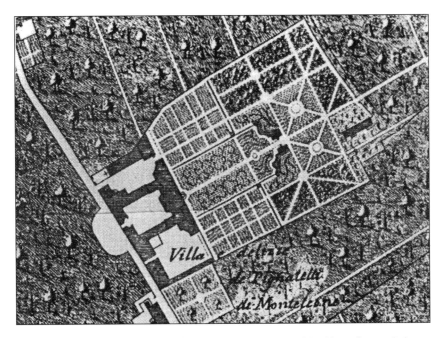

Figure 5.4 An extract from the Duke of Noja's 1775 map of the Neapolitan and circum-Vesuvian areas: Villa Pignatelli di Monteleone, Barra (1728).

124

passes through the garden of the now derelict Villa Lauro Lancellotti at Portici, within a few metres of the ruins of its baroque coffeehouse. Indeed, examples of decay and neglect are common among the villas (Fig. 5.5; Gleijeses 1980).

However, the main influence on the villas has been that of increasing urbanization. More than half a million people now live in the 17 municipalities of the circum-Vesuvian zone (Formica 1966, Spooner 1984). The average population density is more than 3700 per km^2, whereas at Portici (area of municipality 4.52 km^2) it is more than 18000 per km^2. From 1951 to 1971 the population of the area grew by 40 per cent, while the number of dwellings grew by 57 per cent and now averages more than 500 per km^2 (Table 5.1). The municipalities have been very slow to adopt and enforce planning controls. Several communities (such as Torre del Greco and San Giuseppe Vesuviano) are expanding into the main volcanic hazard zones, including future lava tracks and the predicted location of flank eruptions (Scandone 1977, Barberi et al. 1983, Alexander 1999).

Increases in the local population have transformed both the villas and their parks and gardens: a quarter of the 37 parks and gardens have been destroyed and 12 (8 per cent) have been much reduced in size by the sale of land for building lots. Eighteen are the sites of market gardening and horticultural enterprises. Of the buildings themselves, only 15 (10 per cent) remain the private homes of individual owners, and 110 (77 per cent) have become multiple-occupancy dwellings for groups of families. Excepting one recently demolished villa and two that are abandoned, the remaining 16 are given over to a variety of civil and military uses (the Royal Palace at Portici is now the Faculty of Agriculture of Naples University).

A public body for the care of the villas was constituted under Italian Law no. 578 of 29 July 1971, the Ente per le Ville Vesuviane. It began work in 1974, governed by a committee of representatives of central, regional, provincial and local government. A decree issued on 15 July 1976 by the Ministry of Cultural Affairs listed 121 of the villas as coming under the jurisdiction of the Ente. The law provided for the restoration and repair of the villas, in collaboration with or substitution of their owners, emergency work on structures and decorative elements, the evaluation of public amenity and architectural patrimony, and a series of studies and publications.

There have been some notable successes, such as the restoration of Villa Campolieto at Ercolano, and its reopening as an art gallery. However, the legal obligation of the owners and the Ente to restore and maintain the villas is not currently being fulfilled, or even in most cases aided in any way, as the Ente has been starved of funds since it was founded and hence has a very limited capacity to fulfil its statutory duties. Moreover, 21 villas (and 39 other historic buildings in the vicinity) do not fall under the jurisdiction of the Ente at all, and several have been demolished (Ente per le Ville Vesuviane 1981).

Figure 5.5 Three examples of decay and neglect among the Vesuvian villas. (a) The abandoned coffeehouse of Villa Lauro Lancellotti at Portici. It was once situated in a gracious pleasure garden and is now about 2 m from the main Naples–Calabria railway line; (b) Villa Pignatelli di Monteleone, Barra (1728): decay of stucco on the main façade; (c) Villa Menna, Portici (1742): damage to an eighteenth-century fresco.

In summary, it is as well to remember that the momentous social and economic changes that have occurred in the Neapolitan area are not the only factors to have affected the local architectural patrimony. Many of the villas are palimpsests of pre-existing buildings, which have been altered and added to, especially those of the seventeenth century. Others were altered after their construction during the eighteenth century. As many villas were built rapidly of rough-hewn Neapolitan yellow tuff covered with stucco, decay often set in immediately. Deterioration of the fabric of the villas is not a new phenomenon, but dramatic changes in their tenure and the urbanization of their surrounds have occurred since the 1940s (De Seta et al. 1980). The process quickened in the 1980s with the influx of reconstruction money following upon the earthquake of 23 November 1980 (Alexander 1984).

This leads us to the present day. The modern renaissance of Naples, if that is what it truly is, does not extend to its periphery, where urban decline is still in full swing. To the casual visitor – and the fresh perspective that such a person may bring is not necessarily less valid than that of the habitué – the plight of the Vesuvian villas resembles a visual metaphor for the decline of modern civilization. More even than Lisbon, which still bears the scars of the 1755 earthquake, Naples is the Western world's conundrum. Buildings of a mind-boggling refinement and elegance loom out of an equally mind-boggling squalor (Di Stefano et al. 1967) and are treated by their users with a cavalier nonchalance sufficient to give one an apoplectic fit.[8] Indeed, when AC Napoli won the national soccer shield, one of the villas was spontaneously repainted in its colours, sky blue and white, by the inhabitants (at a time when an adoring populace put the errant footballer "San" Diego Maradona, once star of Neapolitan soccer, on the same sort of pedestal as the Roman bishop St Jannarus, the patron saint of Naples). And all this may one day disappear in a cloud of volcanic ash, to be reverently disinterred by the archaeologists of the fourth millennium!

The sense of precariousness is underlined by a map issued by the Italian National Research Council's Vesuvian Observatory (Barberi et al. 1983, Dobran et al. 1994). In the event of a major eruption, three hazard zones would form around the volcano. In zone C, ashfalls would pose a danger to human respiration and to roofs, which might collapse under the accumulated weight. In zone B, victims would be burned, possibly by inhaling hot gases and ash. But anyone left behind in zone A would be carbonized or vaporized (a modified hazard and risk map is shown in Fig. 5.6). The plastercasts of victims from Pompeii and Herculaneum provide a mute testimony to the agonies that would

8. See Guadagno (1971) and Leone (1991); for an English-language account, see Belmonte (1989).

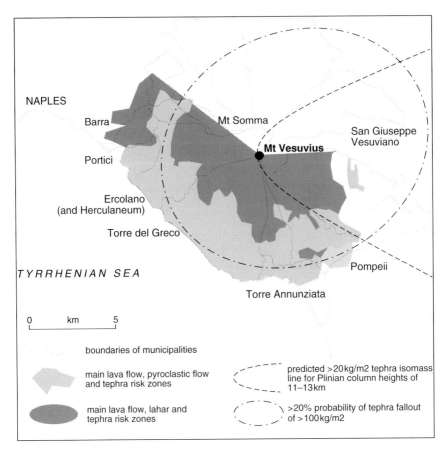

Figure 5.6 Volcanic hazard and risk map for the circum-Vesuvian area (various sources).

ensue if evacuation failed (Sigurdsson et al. 1985, Carey & Sigurdsson 1987).

But such facile explanations should not distract from the progress that has been made in architectural conservation and civil protection. It is more reasonable to view the matter in terms of the old Aristotelean notions of *generatio* and *corruptio* (Aristotle 1950 edition), creation and destruction. More than merely complementary, things must be destroyed if they are to be created. To some extent, then, hope lies in the damage wrought by natural hazards, although we all have a deep moral obligation to ensure that the *generatio* is worthy of our creativity and is not corrupted by its very worthlessness. No time seems more fertile for such considerations than the advent of the third millennium, but that time of renewal and retrospect should not be entered into without due consideration of whether the millennial concept has any real meaning at all. In the next section we therefore turn from the past to the immediate future.

Millennialism

Shortly before new year's day 1998, a national daily paper carried a short article[9] that explained that virtually all the major predictions made by horoscope writers, magicians and crystal-ball gazers for the previous twelve months had not come true. Of the year's most salient events, only the most blatantly predictable had been foreseen. To one who loathes astrology, this afforded a degree of "I-told-you-so" satisfaction, especially as the newspaper that carried the article had periodically, faithfully and quite seriously reported all the predictions in question. A victory for science, perhaps, although not over credulity, as hordes of people continue to yearn for mystical solutions to life's conundrums. A temporary, perhaps pyrrhic, victory for objectivity, then. But it also introduces a wider question regarding time and the future.

Throughout the world, human imagination has been captured by the year 2000. No matter that calendar time is a purely artificial construction, the fruit of Europe's medieval monasticism, which enables us to mark as discrete an abstract concept that has no internal divisions and no beginning or end that we are able to visualize. No matter that the starting point of the modern calendar is, in the minds of most of the world's citizens (that is, the non-Christian majority), entirely arbitrary, and therefore so is the starting point for the third millennium. No matter that there is equivocation about whether the calendar started at zero or one, and whether the Son of God was born at the start of the first year, or in 4 BC or AD 6. All four digits of the odometer of time flip over in the year 2000 and that is a Significant Event for all humanity, not merely for computers that cannot cope with the zeros of the date.

But what does this Significant Event actually mean? Premonitory signs in the late 1990s and similar events from history suggest the following. Like so much of human experience, millennialism (or, if one prefers, "millennium fever") is composed of both positive and negative emotions. The former include a sense of optimism and renewal, a gigantic new year's resolution to make the world and the human race better and to bestow that initiative to the generations of the third millennium. The negative emotions cluster around well known anxieties. Perhaps only a small lunatic fringe believe that the millennium will precipitate the end of the world and that we shall all be consumed in the fires of wrath and retribution. But foreboding is much more widespread. As discussed elsewhere in this book, catastrophism has come back into fashion and initiative has been attenuated by a general failure of nerve.[10] Doubtless some of the latter represents justifiable prudence – not wishing to tempt fate by displays of opulence

---

9. *La Nazione* (Florence), 28 December 1997.
10. Clark (1953) saw signs of this at another significant point in history, the Renaissance.

when the empire is about to crash down into ruins and the barbarians are battering at the gates of the city. This is further complicated by the fact that the modern barbarians emanate from within society, not from outside, and they have required no Trojan horse or any other subterfuge to make their presence felt.

In the field of geophysical hazards, the year 2000 marks the end of the International Decade for Natural Disaster Reduction (Holland 1989). There should have been decreases in mortality, destruction and damage, and increases in security, mitigation and preparedness. Programs for monitoring hazards, educating the public, training workers and reducing vulnerability should be in place. Permanent structures and procedures for coping with natural hazards should have been created (US National Research Council 1987; Smith 1996: 342–6). At the time of writing, it is still too early to evaluate efforts in this direction, but the final verdict will almost certainly be one of partial success, modest advances made, limited goals met, much more work to do, new and renewed challenges, and so on.[11] As intended by the Decade's prime movers, the United Nations Organization has been the catalyst of much of the progress made to date. As it is unlikely that this role will devolve on individual countries or any other international organization, future developments may to some extent depend on the success of efforts to reform the UN. The need is probably more urgent in the central structure than it is in the constituent organizations, many of which function very well indeed. But the UN's role in the international balance of power and in resolving conflicts around the world is likely to have a considerable impact on the mitigation of complex emergencies, and the development initiatives that are necessary for this to be accomplished. The turn of the millennium is a period in which the balance of power is shifting in ways that are hard to predict. Thus, the future of the UN's basic structure is also difficult to ascertain. This makes it particularly hard to predict the opportunities for development of its secondary institutions, such as the Office for the Coordination of Humanitarian Affairs (formerly UNDRO, the Disaster Relief Coordinator's Office and the UN Department of Humanitarian Affairs).

Whatever its attractions may be, futurology is a weak science, a risky form of intellectual tightrope-walking whose only mitigating circumstance is that one's predictions are likely to be forgotten before there is any chance to demonstrate their accuracy. Hence, to arrive at a prediction about the future of disaster mitigation we must proceed with caution. First, we should note that human reactions to the millennium are fuelled by a classic tension of opposites: the paralysis induced by nervous foreboding against the vigour created by a sense of renewal. In the field of disasters these might easily cancel each other out and lead to a failure to develop the necessary mitigation measures. More

---

11. A more detailed critique can be found in Tobin & Montz (1997: 342–6).

likely, the force of circumstance will override such considerations. Steeply rising losses and sudden, heavy tolls of casualties will induce draconian efforts to mitigate and these will overcome the inertia of faint-hearted decision makers.

What is harder to foresee is the effect of polarization: steeply rising population figures and densities, marginalization and economic instability in the South, technological dependence and capital accumulation in the North. A new tension of opposites emerges, encouraged by the inadequacy of capital and technology transfers. Two broad strategies of disaster mitigation develop, where there should be one. Sadly, the realization that socio-economic development is an equal right for all humanity has not led to its equal diffusion. The greatest challenge therefore materializes as a need to work towards convergent processes of disaster mitigation, tailored to local needs but offering more equal opportunities world wide. Once again, it is probable that the chronic imbalances to the global economy caused by appalling levels of inequality will force the situation, rather than any conscious attempt to remedy it before that stage is reached. Let us not forget that the world has refused to accept a tax of a mere 0.25 per cent on international capital transfers in favour of debt relief and development initiatives in the developing countries. Equally, the richer countries are nowhere near to meeting the 0.7 per cent of GNP target in their development aid budgets (*The Ecologist* 1992). From this, we must conclude that the millennium is not proving to be an occasion for the giving (and the graceful acceptance) of charity and for economic reconciliation. But to study disasters is to acquire a reputation for pessimism, given the human race's persistent inability to reduce the toll of casualties and destruction.

To counteract this, one can only hope for a revival of the spirit that built the great cathedrals, not that which enslaved tribal societies.

Optimism is also required in academic study, which in this field has tended to lose its sense of innovation in recent years. In order to end this chapter on a positive note, the next section will integrate some of the concepts presented in the earlier chapter on methodology and in the subsequent ones on social and cultural factors. This can best be achieved by illustrating the methods through a brief example, which will demonstrate a method that, by looking at a single event phenomenologically from diverse perspectives, is intended to broaden and deepen our knowledge of it. This should also help to show that the study of disasters ought to free itself from the constraints of academic disciplines, even of cross-disciplinary communication, and start to look at phenomena without artificially subdividing them into categories connected with each disciplinary perspective. An additional advantage may be derived from this, in the sense of immediacy and realism that such a technique embodies. Moreover, it has a powerful antecedent in the narrative approach to disasters, where events are described and analyzed according to the temporal sequence of their occurrence, which allows one to visualize the way in which complex realities unfold dynamically.

The holistic approach to disasters: an example

There is a need to return to a more integrated, holistic approach to disasters. Academic training encourages the division of a complex and multifarious reality into its constituent parts in order to understand it better. But thereby one loses the sense of the whole. Moreover, division into disciplines is artificial because it tends to reflect the historical organization of teaching and research, rather than the demands of the problem being studied. The particular character of disasters makes them especially suited to a *non*disciplinary, rather than an *inter*disciplinary approach, for they offer the student a series of imperatives that have nothing to do with academic contemplation. Yet, so specialized has our training become that we find it difficult to approach the subject of catastrophe impartially. Perhaps, indeed, it is impossible to acquire the breadth of knowledge needed to be able to analyze disaster by comparing all of its significant aspects.

However, some modest steps can be taken. These involve a primary choice between analyzing an event in the context of the larger reality to which it contributes and analyzing it in more detail in its own right. In both cases the characteristics of the event itself should be allowed to determine the most important problems to be studied and to dictate the methodology used. We shall now consider an example of the latter approach by examining a concise portrait of disaster drawn directly after it had occurred. The example in question is taken from my diaries, as written during the aftermath of the Campania–Basilicata earthquake of 23 November 1980 (magnitude 6.8), in southern Italy, in which I was directly involved in the role of survivor made homeless by the disaster (Alexander 1981, 1982b, 1990). And so we return to Naples in order to consider what disaster really felt like on the ground, and how it can be analyzed without quite so much detachment and artificiality as is so often the case with academic specialists.

Soldiers, policemen and ambulance men with masks over their faces watch and some of them cry. The sky is grey and the atmosphere laden with cement dust. A mechanical digger advances into the rubble, picks up a scoopful and retreats. A human arm rolls down the side of the hollow left by the digger. Carefully, the rescuers scrape some dirt away. Among the dust and pieces of cement there is the body of a woman. Her hair is matted with dirt and her legs are already rigid

132

and blackened. The body is wound up into a sheet and taken to one of the ambulances, which is already grimy inside from its previous journeys. Inside another ambulance lies the body of a father whose two children, also dead, still cling to him. The diggers rear up like dinosaurs in a swamp. The area is cordoned off, but there are crowds of soldiers and lines of ambulances waiting for the next victim to be discovered under the massive dome-shape heap of rubble that stands at the corner of Via Stadera in the Naples district of Poggioreale.

In April 1950 the construction of three large apartment blocks began on this spot. They were completed in August 1951 at a cost equivalent to US$41 000 each (at the prices prevailing at the time), and stood nine storeys high (34 m). There was some imbroglio over licensing these buildings for occupation, but nevertheless they were quickly occupied, and each housed 20 families. Local talk had it that they swayed alarmingly when the wind blew, but in almost 30 years none fell down.

At the time of the earthquake, most families in the block that collapsed were at home. A christening party and a confirmation party were both in progress, and several families had guests. The earthquake allowed terror-stricken families enough time to run for the stairway, but then collapse was total and instantaneous. Fifty-eight people lost their lives and it took three days and three nights to extract the bodies. The huge mound of concrete, twisted steel, wood and broken furniture was all that remained of 20 homes. Survivors covered the ruins with flowers. Close by, the other two blocks stood empty and forlorn. People lacked the courage, or rather had more sense, than to re-enter. Washing still hung on the balconies and in some rooms a light still burned. One of the blocks was leaning precariously.

The survivors, those who were bereaved, refused a public funeral for the victims and demanded a full-scale enquiry into the tragedy. Promises and pledges were made, but nothing was done. The homeless of Via Stadera were next seen in the streets protesting that the authorities had not even granted them proper shelter. Those who were able to speak of their experiences gave vent to a fury of accusations, mingling grief, rage and shame. In Italy one of the words for tragedy is "disgrace" – *disgrazia*.

Any attempt to bring out the salient characteristics of the above narrative must take account of some limitations. In the first place, this small portrait does not deal with the long-term aftermath, although it does include some information on the history antecedent to the event. Secondly, the catastrophe must necessarily be viewed with hindsight, which is not how it appeared at the time, not even in the days of the search-and-rescue operations. However, various observations on the collapse at Poggioreale will serve to clarify its significance to the wider issues of interpreting disaster.

To begin with, the choice of Naples was not entirely casual. No European city

133

exhibits such startling contrasts, and none has had such varied social and economic fortunes, often at the same time (Leone 1991). It is a city of paradox, where mind-boggling richness assaults the eye in the same field of vision as the deepest refinements of taste, the crassest forms of suburban materialism, and the most brutal forms of inner-city decay, all against a setting of precarious volcanic hazards and Europe's highest population densities by far (Portici, near Poggioreale, reaches 18 000 inhabitants per km^2, nearly four times the density in central Milan). After centuries of maladministration, corrupt government and organized crime, Naples emerges as a metaphor for the decline of Western civilization, a new capital of the baroque, one in which the tension of opposites is between the bad and the worse (although in its defence I should add that in the mid-1990s Mayor Antonio Bassolino made gigantic strides in reversing the trend). The richness of Naples lies in its sheer heterogeneity, and that makes it an ideal laboratory for whoever is strong enough to study it closely (Guadagno 1971).

Moderate to large earthquakes often cause structural collapse that is highly localized within large urban areas. Good examples can be found from Mexico City in 1985 (Esteva 1988), Manzanillo (Mexico) in 1994, Sakhalin (Russia) in 1995 (Alexander 1996, Porfiriev 1997) and so on. In the municipality of Naples in November 1980 there were only two sites of major structural collapse, although minor and partial collapses were more widespread, including the side of the building in which I had been staying on the day before the earthquake. Poggioreale is a poor quarter of the city, part of a 1950s extension of shoddy and ill planned development that succeeded in extending to the north of the inner-city the decay that afflicted the Spanish Quarters, the heart of the old city. Moreover, Naples has only recently begun the process of geographically segregating its urban functions, and hence, in 1980, residential development coexisted with artisan's workshops, commercial premises, light and heavy industry, chronic air pollution and hazardous installations (Spooner 1984). Industrial explosions are periodic events even to this day. In such an atmosphere of precariousness and congestion, and given the daily problems of crime and unemployment (which produce an unholy symbiosis, as criminal syndicates are great providers of work to the unemployed), people live on vulnerability, become accustomed to it, expect no better and see no ways to reduce it. The result is a culture of passive acceptance that allows no alternatives and forces people to the bottom of the socio-economic scale.

Despite this situation, enquiry and recrimination were necessary parts of the social reaction to such a disaster. So should prosecution be, although in this case the habitual paralysis of the Naples city government during the 1980s, the inefficiency of the justice system and the hierarchy of political patronage all conspired to stultify any such initiatives. Moreover, the passage of time and the almost miraculous survival of other highly vulnerable buildings tended to obscure the original responsibilities for the tragedy. Yet recrimination is as necessary to society as grieving is to the bereaved. Together, the two forces make a powerful admixture and one that can motivate individuals to seek redress, even in vain, for years (Neal 1984, Horlick-Jones 1996). It is a pity that such motivation can so seldom be used constructively as a means of opening an avenue to greater security in the future.

The instantaneous structural collapse of a large occupied building poses very difficult technical problems of urban heavy search-and-rescue (USAR) and of the recovery of bodies (Olson & Olson 1987, Krimgold 1989). Slabs of concrete and steel beams cannot be moved easily, not merely because of their weight and the precariousness of piles of rubble, but also for fear of damaging the survivors or bodies trapped beneath. Although much expertise has been accumulated on techniques of USAR, and specialized apparatus exists to locate and extract survivors, it is not uniformly applied. In USAR as in other aspects of disaster management, experience is a great teacher, but one that often metes out its lessons in a negative way. Thus, equipment and training tend to be acquired *after* the need for it has been demonstrated in some disaster that has elicited a suboptimal response from the rescue teams. Fortunately, however, the globalization of relief and increasing use of the Internet have begun to spread a more uniform, universal culture of professionalism and sharing of experience in this field,[12] although there still is a long way to go. Special attention needs to be given to ways of tackling *critical incident stress*, a subject of post-traumatic stress disorder, in which emergency workers are deeply affected by the apocalyptic vision of death and destruction that they have had to deal with. Although in Anglo-Saxon cultures there has been a fair amount of research on the subject,[13] it is not clear what scope there is for devising more universal strategies for dealing with the problem, especially with reference to emergency workers from other cultures.

In one respect a disaster such as that at Poggioreale can be viewed in terms of a temporal imbalance of resources. The contrast is striking between the lack of resources devoted to the safety of the three buildings before the earthquake and the plethora of emergency services that were on site after the event. This is indicative of what we might call the central **paradox of mitigation**: it is generally more efficient and cheaper to devote money and attention to protection before disaster than to spend similar amounts on emergencies and damage afterwards (Anderson 1991). However, without hindsight or a detailed feasibility study, there is little to stimulate the comparison, and, moreover, post-disaster expenditure cannot be avoided, even if it can be reduced by prior mitigation.

---

12. For example, see the American Rescue Team's web site at:
    http://www.Acosta.com/AmerRescue.html.
13. See Shore (1986) and Brown & Campbell (1991); an example is given in Nolen-Hoekesema & Morrow (1991).

A few streets in the old quarters at the heart of Naples have permanent signs that warn of the danger of structural collapse. Inner-city decay, so often held to be the bane of the late twentieth century, has existed there for centuries. Architectural precariousness spills over into a form of cultural vertigo in which the nonchalance with which risks are assumed daily boggles the mind. Years before the 1980 earthquake I knew an old lady who lived in the Spanish Quarters at the heart of the city and would not take her afternoon nap without placing her shoes and handbag by her side, "in case the earthquake comes". It was an almost hereditary precaution, a reaction conditioned by events from the distant past and kept alive by the ebb and flow of anxiety. Each time that cholera threatens to break out or the ground trembles, street-corner shrines are dusted down and painted up. In the meantime they are neglected. But one cannot hope to understand the social reaction to disaster without knowing the cultural underpinnings that lend meaning to the people's apparent illogical actions and reactions (Anderson 1967, Brislin 1980, Gherardi 1998). Instinct, nonetheless, tends to create a predominantly negative culture of hazard and disaster. It must be externalized and made more concrete before it can be used to constructive advantage in favour of civil protection.

In synthesis, victims, survivors, rescue workers, journalists, politicians, apartment-block builders and passers-by all participate in the same reality in mutually reciprocal roles that collectively define a residual level of tolerance of risk and hazard. Events accumulate in history, society evolves under internal and external pressure, tolerance levels alter with society's consensus, but the prevailing fundamental culture of hazard slows the pace of change and reduces it to small increments, surcharged by each new disaster that occurs. To utilize the post-disaster "window of opportunity" in favour of better mitigation, one must recognize the cultural constraints on action, gauge the level and direction of societal consensus, and then act within the realm of what will be tolerated and supported by the beneficiaries. This requires a holistic attitude to the analysis of risk.

Having examined the social, cultural and historical underpinnings of human responses to disaster, we will now move on to consider one very pertinent aspect of society under the duress of natural hazards: its relationship to technology and technocracy. This was touched upon in Chapter 3 (e.g. see Fig. 3.1), and in Chapter 4 (see Fig. 4.1), but it deserves a more extensive and systematic treatment.

CHAPTER 6

Technology, economics and logistics

The world's affluent societies, and even to some extent the poorer ones, have entered a technological age. A quiet but fundamental revolution has occurred on the basis of the electric motor, the internal combustion engine, the cathode ray tube, the transistor, the computer microchip and the satellite. Technology has become progressively cheaper, more widely available and more sophisticated. The price of this revolution is undoubtedly very high but presently incalculable, for the consequences have yet to be assessed fully. For the time being, human society has learned to embrace technology but not fully to assimilate it. The cultural shock has been followed by cultural disorientation, for life's points of reference are being changed profoundly by technological development. In this chapter we will examine the technological changes that have occurred and assess their impact on perception of and attitudes to disasters.

The power of the mass media

Nothing illustrates the perils and challenges of the technological revolution better than television. It has assumed not merely the position of the principal medium of electronic communication but also the status of an authority that is difficult to challenge. My own experience of appearing in and making television programs, limited though it is, suggests that the process of orchestrating facts and opinions into a form acceptable to television leads them to undergo subtle and often profound changes. In short, television distorts reality by abstracting it: at best this renders things more comprehensible, at worst it can turn truth into lies.[1] Yet as a medium it is inescapable: not only has it reached practically all homes in the affluent West but it has also invaded public space – shops, airport

---

1. See Anderson (1997) for a close analysis of how it does this with respect to environmental extremes.

lounges, aircraft, buses and many other places. Power is the control of what appears on television – on the popular channels, at least – and even those people who would despise it have little option but to treat the medium with respect. Television thus has the ability to inform people about hazards or keep them in the dark, to tell great truths about natural catastrophes or to perpetuate myths, and to motivate public solidarity in the face of the suffering that disasters cause, or to not stimulate it. No matter how civil protection is organized, television is a part of it.

At the end of the twentieth century, mass media of all kinds are susceptible to three distinct problems. First, the filtration of news, which is widely practised in order to make it palatable, involves an excessive degree of selectivity. In order to avoid too heavy a demand on the viewer's concentration, news is reduced to "bite-size chunks". That which lacks immediacy and instant relevance is discarded. The result often bears little relationship to the distribution of important events around the world. Secondly, the process of imparting an angle, or slant, to news is a dangerous one, which easily leads into the trap of confusing fact with fiction (Goltz 1984, Singer & Endreny 1993). This is especially the case when news merges into entertainment. Indeed, the latter has begun to assume an authority that amounts to the status of an icon and it reflects very uneasily on popular priorities. A complete outsider subjected to a prolonged dose of popular media would probably assume that entertainment is far more important in our lives than news, current affairs or education, and that it is the principal source of our values. The task of piecing together some form of objective reality from the modern media is laborious (Anderson 1997). Given the plethora of sources, it is not impossible, but it involves a great deal of reading between the lines, interpreting the media's way of presenting things and searching for information in obscure places. Few training opportunities exist for this difficult enterprise, and most people seem to prefer to have their news digested for them. Making sense of disaster via television images is therefore not the simple exercise that it purports to be, but a difficult and highly specialized enterprise.

The role of the mass media has been studied in a wide variety of situations, including mass emergencies and disasters, although largely with respect to newspapers (which are easier to monitor than television or radio) and almost exclusively with respect to Western media sources (Alexander 1980, 1997, McKay 1983). Many of the findings cast considerable doubt on the ability of news personnel to report on disasters in a rational and objective manner (Seydlitz et al. 1994, Ploughman 1997). Although it is still too early to make a very comprehensive and systematic analysis of the discrepancies between fact and reporting, some consistent regularities have emerged. For instance, there is often a wide difference between reporting of domestic and foreign disasters, in terms of extent of coverage, aspects emphasized and reporters' attitudes to the events and people they are describing (Alexander 1980, Needham 1986). However, a more positive school of thought also exists in which it is held that careful briefing and management of the news media can enhance their role in providing useful information to the public in emergencies (Scanlon et al. 1985).

Television and relief appeals

Television is well adapted to bring home the immediacy of disaster. Graphic images of casualties, destruction and the violence of nature lend a strong sense of reality to events that otherwise would usually be remote in the minds of the public. But the camera work is not objective. To begin with, there is an understandable tendency to concentrate on the worst, most cataclysmic scenes. It is easy for the viewer to believe that the whole disaster area is in such a state. For example, moderate earthquakes that affect cities often cause spectacular but highly localized structural collapse, perhaps involving only a handful of buildings. Given the tendency of commentators to use superlatives (if that is not too positive a word) when describing damage, the viewer who sees images of a large building that has collapsed, perhaps with victims trapped inside, is easily convinced that the whole city has been shaken to the ground. Sober reality is very much less exciting than what can be made of an eye-catching image.

Nevertheless, nothing can motivate a public response to disaster quite as effectively as television coverage. For example, there was a telling image from the 23 November 1980 earthquake in southern Italy, one in which damage was indeed widespread in at least 36 urban areas. In a devastated town centre only one building remained functional amid the rubble. It was a bar and it had a television set on the wall and still had electricity. A group of local men sat huddled around the flickering screen. They were watching live coverage of a devastated town centre strewn with rubble.[2]

When heart-rending and catastrophic scenes appear on television, there is often a substantial public response in terms of solidarity, donations, and perhaps volunteerism (Moran et al. 1992). Were such a thing possible, it would be very interesting to attempt to measure the degree of correlation between the duration and magnitude of television coverage and the longevity and strength of public response to a disaster. The fundamental hypotheses would be first that, by covering or not covering the event, television is virtually capable of turning public response on or off like a tap, and, secondly, that the nature of television coverage bears a positive relationship to the strength of public involvement or donation.[3]

---

2. I am indebted to Dr Robin Stephenson for this example.

Most probably, no objective means exist to test these hypotheses – there are too many extraneous variables to monitor. However, there are superficial indications that the model is valid to some extent. In the West, relief appeals seem to last about as long as television coverage. In the case of domestic disasters and ones that occur in neighbouring states, both may be prolonged over months. On the other hand, in the reaction in Europe and North America to the 1988 Armenian and 1993 Maharashtra (Indian) earthquakes, television coverage showed scenes of catastrophic structural collapse and deep misery and helplessness among survivors. The relief appeal was intense but was as brief as the airspace devoted to these two events.

No clear evidence exists as to whether government-sponsored relief is directly motivated by televised coverage of disaster areas. One hopes that it responds to more rational and objective demands, such as the precisely for-matted relief appeals coordinated and disseminated by the United Nations Office for the Coordination of Humanitarian Affairs. But there is certainly an indirect effect of television, as public reaction to disasters can easily result in a change of political response, usually towards increased generosity or greater alacrity. On the other hand, it is probable that relentless television coverage may result in "donor fatigue", at least among the public, if not on the part of the authorities. In this, donations diminish under the pervasive assumption that more help can do little good, in other words, that the problems cannot be solved by appeals for extra cash (Hendrickson 1998). In many cases the problem is so politically, socially and economically complex that this is both truth and false: the money is useful only if it is spent in appropriate ways (Kirkby et al. 1997).

Television is, by and large, neither equipped nor accustomed to search for the answers to such questions. Indeed, it often ends up by reducing complex realities to the level of inelegant and unrealistic oversimplifications. There seems to be an unwritten rule not to concentrate too long on any particular feature or interview. Thus, television makes headway by virtue of its ability to change scene instantly and frequently. Whereas initially this merely lent it a vivacity that other media lacked, the pace of scene-changing has speeded up to ridiculous levels. Modern television is often breathlessly frenetic, a flickering pageant of filmclips that, at their worst, last less than a second each. The master of the medium is he or she who can compress the greatest degree of apparent wisdom into the sound bite, the one-phrase comment uttered on leaving a build-ing or getting into a gleaming chauffeur-driven car. Rather than illuminating the world, we are descending into sloganeering, in which we derive our comfort from *reductio ad absurdum* (which is unforgivable mental laziness), not from proper knowledge of complex situations. This is particularly worrying in the

---

3. Compare Ingram (1977) with Benthall (1993).

context of disasters, as they tend to be raggedly uncertain events, at least in their emerging phases. Uncertainty and the sound bite are irreconcilable: although one can summarize an indeterminate situation by saying "we just don't know", to make it credible the sound bite requires a much more definitive closure.

Voyeurism

One of the least palatable aspects of television coverage of disasters (and of many other aspects of human misery) is its tendency towards voyeurism. The analogy with sexual transgression is not my own invention, for others have written in passionate terms of "disaster pornography" (Omaar & de Waal 1992). In both still photography and video the camera intrudes into moments of intensely private human suffering: the survivor in front of her ruined home, the father who discovers the body of his child amid the rubble, the church or mosque that collapses onto a group of worshippers. One of the most memorable of these cases involved the impact of the Nevado del Ruiz lahar (volcanic mud-flow) on Armero, Columbia, in November, 1985. At least 22 000 people died as the lahar swept late at night through the town, the inhabitants of which had not been adequately warned (Voight 1990). The following morning, cameramen were able to photograph living victims trapped up to their necks in soft volcanic mud and perhaps conscious, but who could not be rescued.

However, nothing is more eye-catching to the television viewer or magazine reader than images of starvation, malnutrition, disease and suffering. In both television and the print media the effect is often heightened by the way in which such images are shamelessly juxtaposed with advertisements for luxury goods – fast cars, expensive consumer durables, exotic holidays and fattening foods. After decades of this it is striking, but no longer remarkable, how little protest such contrasts elicit.

Despite my earlier ruminations on the impact of television on relief appeals, much "disaster pornography" has little appreciable long-term effect on public reaction to disaster. The viewer or reader is fascinated by the images of human suffering, and perhaps fleetingly thankful not be part of it, but that is all. I doubt whether international volunteerism is really motivated by images of starvation or injury, as these form no real inspiration to positive action (Wolensky 1979, Moran et al. 1992). The reason for the ineffectiveness of such images is that they are often as isolated as they are fleeting. We hear that 500 Chinese people have died in a flood and we see a brief image of washed-up dead bodies, but that is all; it hardly affects us on a personal level.

Telecommunications technology and institutions

Let us now make a short excursion into the origins, and some further implications, of telecommunications technology. It is easy to forget that the impetus of technological development has so often been military. This is true of the original reasons for building freeways, motorways and autobahns. It is true of forms of telecommunication and satellites, and it is true of the Internet. Military technological development is a riotously expensive never-ending quest for prowess and might. Besides questioning its moral status, we may also ask whether the best way to develop technology for peaceful uses is to build on military research and development when this becomes sufficiently obsolete to be declassified. What does it say about society that some of the richest and most influential technological universities – MIT and Johns Hopkins, for example – have derived large proportions of their incomes from military research? What are the implications for efficient development of peaceful technology in the billions of dollars invested in the "Star Wars" space shield and such projects of immense strategic futility? What are the spin-offs of military technology for civil protection against disasters, given that there has always had an uneasy relationship between the civil and military approaches to managing disasters (Anderson 1969). The answer is that the development of technology for bellicose ends has offered remarkably few direct benefits to civil protection.

Nevertheless, whatever its origin, telecommunications technology has blossomed in the 1990s as never before, and has also burgeoned in the fields of disaster prevention and emergency management. The constant annual expansion of Internet use and connectivity, the impressive array of telecommunication satellites, falling real prices for hardware, software and access, remarkable plans to connect telephone, computer and television – all could not have been accurately forecast a decade ago.[4] Even less easy to predict is the revolution in public attitudes and predilections that information and telecommunications technology will engender. Curiously, a prophetic if rather apocalyptic view of this appeared in the early twentieth century in E. M. Forster's one excursion into science fiction, a short story entitled *The machine stops*. In this he described the process of intellectual, physiological and ecological atrophy that afflicts a world that has become entirely dependent on a technology that generates and repairs itself. His proposed solution involved a reinvigorating escape back to a primitive form of naturism, a proposal that would certainly worry many of the enthusiastic protagonists of our "brave new world". Indeed, for an alternative approach one might go back to the Vorticist and Constructionist movements of the 1910s and 1920s, which held that new technology had to be embraced and

---

4. For example, compare Carroll (1983) and Mitchell (1997).

converted to benign uses rather than eschewed. This age-old dilemma remains: do we accept or reject the challenge posed by technology? How can we assimilate it without losing our identity and our reason?

Another growth field is that of information management. In 1996 US government agencies spent US$4 billion in collecting GIS data, one fifth of their total outlay for the acquisition of data. Intense interest in the use of GIS in disaster management relates to three main endeavours: the development of a comprehensive database on hazards and the built environment, the analysis of cost and benefit for mitigation projects, and opportunities to integrate GIS with other technologies, particularly those connected with remote sensing, monitoring, and emergency communications. The methodology is particularly well suited to the rapid definition of the scope of a disaster and potentially to real-time monitoring of impacts and relief efforts (Hodgson & Palm 1992, Watson 1992). Assumptions and criteria used in the selection and collection of data can determine the outcome of the GIS analysis. Although the use of GIS may facilitate intuition, principally by displaying information in mapped form and by demonstrating information by overlay, it is no substitute for intuition.

The Internet and disasters

One of the most remarkable vehicles of technological change is undoubtedly the Internet, the computer "network of networks", by which messages and information can be exchanged with extreme rapidity and efficiency. Because it is beginning to demonstrate immense potential in the field of disaster management and mitigation (Gruntfest & Weber 1998), we will now examine the growth of the Internet in detail, for both the potential and the achievements in this field demand a critical appraisal.

Internet connectivity extends throughout the developed world and to places as far afield as Greenland and Antarctica. It reaches Fiji and Bangladesh; it brings Australia into close contact with Canada, South Africa closer to Russia. But the question of access and connectivity remains thorny. Large parts of Africa currently have no Internet connectivity at all. Even where there is some, as in Bangladesh, it may be severely limited to but a few major institutions, each with no more than a handful of workstations. Although there are more than 100 million users, that still leaves nearly 6 billion without access and, given the level and trends of world poverty, they will remain the overwhelming majority – the technologically deprived (although plans to socialize Internet access in rural communities may offset this in a limited way). Among those people who have access, there are large discrepancies in functionality, speed and price, depending on factors such as the quality of uplinks and telephone lines, the cost of telephone tariffs, and the power of personal computers and modems. Hence, we would be wrong to think of the Internet either now or in the foreseeable future as a universal or equitably distributed resource.[5]

Once one has access to the Internet, there are questions of content and the quality and utility of information.[6] In some respects the current social organization of the Net resembles the utopian political view of the anarchists. It is held that no one is in charge, that once access has been granted one is free to develop, display and offer to others one's ideas, whatever they are. Censorship is limited by the topological complexity of the Internet and by the difficulty of establishing jurisdiction over something that is so heavily internationalized. It is a global village, a worldwide tower of Babel, even if it is a rather exclusive one.

By virtue of the rapidity and flexibility with which the Internet transmits information, and the opportunity to participate in truly global efforts, many emergency managers have high hopes for its future role as a catalyst for mitigating the impacts of disaster (Gruntfest & Weber 1998). There is no doubt that it gives a sense of immediacy, and several other practical benefits, to the field. It has undoubtedly helped disseminate useful information (disaster plans, data on emergencies, manuals, and so on) more widely than would otherwise have been the case. It has also contributed to the internationalization of the debate on how best to tackle disasters (Keller et al. 1996). But in reality, there are drawbacks and limitations to the apparently free circulation of information. To some extent, not only the character of the network but what it is able to purvey is shaped by the providers of service. Well meaning attempts have been made to curb some of the excesses of pornography and extremism using national legislation, although with checkered success. On a more sinister level, there are constant attempts to impose tariffs on the transfer of information (a situation that already exists for many private individual clients and telephone modem users). If these are eventually successful in altering the nature of the service offered to Internet users, they will also end up by changing the character of the data that are transferred.

Another drawback is one that is frequently encountered with manifestations of glorious anarchy. Many of the contributions to the Net are inane and of very low cultural value, and this is particularly evident in some of the discussion groups that have formed to debate emergency-management topics. Poverty of expression abounds (but not the least as a result of the tendency of intellectuals to shun the Internet, or at least to participate in it only passively). Moreover, the

---

5. Despite phenomenal growth, the proportion of the world's population that regularly uses the Internet is still minuscule and is likely to remain so for a long time. So much for mass communications and global coverage.
6. See Quarantelli (1996) for an exceptionally lucid discussion of this problem.

English language is so pre-eminent that it virtually stifles intercourse in other languages (despite attempts to rectify the situation by users in non-anglophone countries). And such is the quality of English as expounded on the "Net" that it frequently descends into incomprehensible or semi-comprehensible collections of jargon, acronyms and slang. Correct grammar goes out the window. This, of course militates against the free transfer of ideas across international boundaries.

Nevertheless, there is now an extraordinary plethora of freely accessible information of all kinds and persuasions (Gruntfest & Weber 1998). Will the excessive emphasis on quantity ever be replaced by a corresponding emphasis on quality? The justification that is often given for this situation (and likewise for plans to offer cable and satellite viewers 500 television stations) is one of *choice*. But choice is a much overrated ideal (Bernal 1949). The net effect of an excessive range of choice is to reduce access and obscure the route to anything that is truly worthwhile. Too much choice blunts the critical faculties and adds to the passivity that consumer culture has succeeded in substituting for curiosity. It also confuses people who are searching for useful information on disaster management on the Internet, although the need spontaneously to develop criteria to evaluate the quality of what is found may well be an unexpected benefit in its own right. Finally, although the range of information sources available on the Internet is vast, the permanence of websites is by no means assured: they can disappear just as they once appeared.

In order to develop a rational perspective on the Internet, one should be wary of making too many comparisons with other means of communication. It has often been argued that the Internet will become a substitute for other media. Yet in my view it should not be regarded as any form of a substitute for printed books and libraries. It is peculiarly adapted to the dissemination of perishable information, although to use an Internet-based newspaper or periodical requires one to develop such a different technique of discrimination in selecting what one wishes to read that it is not entirely comparable with reading a printed newspaper or magazine. In short, for many users the Internet has added a new layer, a new dimension, to a life that was already bristling with complications. But one cannot afford to be a Luddite and hope that the "Information Superhighway" will suddenly close down; at the most it will suffer from ever more frequent and complex go-slows and traffic jams as usage exceeds capacity. This may well pose severe limits on its use for emergency traffic, although these can, at the expense of public participation, be reduced by restricting such messages to **intranets** (internet-like networks within organizations) rather than opening up access to all and sundry. Like highways for vehicles, creation of the infrastructure tends to overstimulate use. The only saving grace is that there are many more easy routes through the Internet than one could find in a road network – such is the beauty of a thoroughly mutable topology.

The Internet also has a theoretical geography, one that stems from earlier technological developments. In 1966 the geographer William W. Bunge wrote that the geography of the world has been irrevocably altered by the nuclear age (Bunge 1966). Strategic weaponry puts us all in the front line and territoriality takes on an entirely new conceptual meaning. Political geography can now only be understood as a cartogram, a topological transformation in which the abstract characteristics, not linear or other real measures, determine the size, shape and pre-eminence of places. In short, the dimensions of countries, at least in conceptual terms, and their reaches are determined by power and aggressiveness. Bunge was writing at the height of what the cultural commentator Jeff Nuttall referred to as "bomb culture", the collective psychosis that emanated from the realization that a few crazy political decisions could end in our annihilation. It was an idea that had profoundly depressing repercussions in many aspects of human endeavour. The psychosis, but perhaps not the threat, has abated and we now face its successor, techno-culture, the machine language of the mind. The possible psychological impact of this may include a tendency to treat disasters in a more detached but less humanitarian way, to make decisions without being involved in their consequences. Whether it turns out this way or not, time will tell.

The topography of the Internet is practically unfathomable. In a matter of minutes, one can log on in Boston, access a site in Sydney, make one's way to New Delhi and from there log on again contemporaneously in Boston. It is a system that demands rational use if it is to have any sense at all. At the level of a superficial analysis, accessing the Internet is to enter the gates of the global village, with the means and the passport to go wherever one wants to. In actual fact, this is something of an illusion. In the first place the number of sites and volume of information that they contain is heavily dominated by the USA, the country that originally devised the Internet. Most of the thousand or so sites related to disasters and emergencies are, in fact, in the USA. The contributions of the other 150 countries do not reflect their cultural or scientific worth.

Secondly, it is paradoxical that mere access to this putative global village does not make the user a true citizen of the world. Parochialism is as abundant on the Internet as it is in any other walk of life. Nevertheless, the geography of the Internet is attractive because it is not based on physical distance, merely on telecommunications connectivity, which at its best is nearly instantaneous.[7] And it is attractive because of the possibilities that it offers, the potential to create new communities of people who would otherwise have little or no opportunity to work together or relate to each other through their common interests. This includes the challenge of sharing more information on disasters and how to

---

7. The imminent arrival of Internet II will apparently speed up connection times by up to two orders of magnitude with respect to the speeds current in the 1990s.

mitigate them, which, most fortunately, is being faced up to in many positive ways by the growing community of emergency managers who use the Internet.

To fulfil the aims given at the beginning of this section, let us now examine what cyberspace has to offer to the student of natural catastrophes and subject it to a critical examination. Currently, hundreds of sites are available on the World Wide Web (Butler 1997) or through file transfer protocol (i.e. ways of accessing information and transferring computer files) that deal with the following:

- information and databanks on events, mitigation techniques and research endeavours
- real-time and near real-time access to information on currently existing events
- exchange of scientific information on hazards, and of practical experience on civil protection, hazard mitigation, search and rescue techniques, and recovery from disaster
- circulation of international relief appeals
- information bulletins and discussion groups on current events regarding relief endeavours, scientific monitoring, scientific data, geophysical phenomena and mitigation measures
- news bulletins, briefings and debriefings on current disasters
- thematic dissemination of news on the impact of particular disasters on, for example, insurance, agriculture and the commodity markets
- emergency plans and policy statements
- photograph and map collections regarding particular disaster events
- consultancy services in which users can query experts
- on-line conferences and instructional courses.

The prospects for developing these services essentially comprise a large potential for greater international collaboration, the opportunity to attain a wider and more comprehensive worldwide view of disasters, and the challenge of creating a truly global network of experts and people who have something to contribute to disaster management and mitigation (Cate 1995). Some aspects of the Internet's potential in this field are already well developed; for instance, its use for launching relief appeals (Anderson 1994),[8] for distance learning, to accumulate data on a worldwide basis, and for the supply to the public of "perishable" information on current events. Other aspects are being vigorously pursued in the more technologically advanced countries, especially attempts to create robust networks for the instantaneous diffusion of emergency information, and measures to keep the mass media informed as events unfold. Still

---

8. In addition, see the Volunteers in Technical Assistance site http://www.vita.org, which carries United Nations, US Geological Survey and NGO situation reports.

other aspects of the Internet's potential have yet to be fulfilled, including its ability to facilitate the coordination and standardization of emergency plans and to help formulate a worldwide strategy for coping with disaster. Likewise, much work remains to be done before the Internet can be used routinely to transmit real-time data as part of the monitoring of remote sites and distant geophysical processes.[9] Another distinct challenge is to link it more closely to remote sensing and geographic information systems for better spatial data depiction (the principal problem here is the amount of data contained in each image). Although some progress has been made in enabling experts who are not directly involved in particular disasters to seek information or offer advice to fieldworkers, this is another sector in which the potential remains largely untapped.

One of the principal objectives is to create truly global working groups on particular themes in disaster studies. In this the volcanologists have been first to take up the challenge and, indeed, they have proved singularly adept at introducing innovation via the Internet.[10] Undoubtedly, this is because of the global nature of volcanism, the remoteness of many scientifically interesting volcanoes and the fact that only a score of the world's 500 active volcanoes are intensively monitored (Tilling 1989, McGuire et al. 1994). The Internet has proved an especially useful means of overcoming these problems and efficiently disseminating volcanological information. Another objective is to be able to add the Internet to other channels of emergency communication, and hence to be able to use computers interactively to manage emergencies (Spurgeon 1996, Gruntfest & Weber 1998). When combined with physically robust means of transferring data (especially using satellite links), the diffuse topology of the Internet should provide enough redundancy and duplication of routes to avoid the problem of blockages and breakages in the system as a result of damage in disasters, although this supposition has yet to be tested fully.

Despite the apparently bright future of Internet use in disasters, there are inevitably drawbacks. To begin with, the plethora of sites is confusing and leads to difficulties of finding, selecting and assimilating information. Various sites have attempted to overcome this problem by offering a thematic list (menu) of access (gateways or links) to other sites. However, this tends to lead to topographic contortions that reduce the efficiency of data transfer and network usage, and to overload the user, as there is quite simply too much information to be able to assimilate it all. It is perhaps a blessing in disguise that not all of it is high quality; indeed, some is evidently on line more for show than for genuine edification. In this respect the incentives to make information freely

---

9. See Alexander (1991c) for details of international geophysical telemetry networks.
10. See the Bulletin of the Global Volcanism Network, which is available via the Smithsonian Institution World Wide Web site (http://www.volcano.si.edu/gvp/), based in Washington DC.

available boil down to the prestige and publicity of having one's compilations looked at by the Internet-using community, as against the drawback of receiving no royalty for the service of making it available (unless one limits access to paying customers). In point of fact, the efficient use of network information demands a new skill of the user: the ability rapidly to assess the quality of what is offered and equally rapidly to obtain and reduce to its essentials the useful material and discard the useless. It is a relentless and fatiguing process. Moreover, special tactics have to be adopted in order to keep track of perishable information (usually that which deals with current disasters). Finally, there are constant problems of equating supply and demand, which lead to frequent overcrowding of the Internet. The US Federal Emergency Management Agency's site logged over two million visitors during Hurricane Bertha in July 1996, which was then an absolute record number, although it no longer is. Inevitably, access becomes slower and data transfer more inefficient when the networks are heavily crowded. This means that, to disaster managers during emergencies, *intranets* – parallel, small-scale versions of the Internet with restricted access – may be more useful than the Internet.

Ability and willingness to invest in the necessary infrastructure will have a critical effect upon the future pattern of Internet use in disasters, and there are already signs of wide discrepancies in ease, efficiency and cost of access. In this respect, political myopia could easily set in, thus restricting the opportunities to innovate and provide for the future. It is salutary to reflect that as late as 1994 it was reported that Chancellor Helmut Kohl of Germany was labouring under the illusion that the "Information Superhighway" was a nominal strip of asphalt. And while private development will undoubtedly provide the demand for national and international facilities, this will not guarantee provision of adequate uplinks, fibre-optic recabling, and the restructuring of telephone tariffs to facilitate digital information transfer. Hence, demand is likely to outstrip supply as the Internet grows, with consequent detrimental effects upon service. This may have serious effects during the period of heavy usage associated with a post-disaster emergency phase.

The next stage is probably going to be the integration of domestic and business technologies into one multimedia system that includes both computer and television. At one level this will provide unparalleled new opportunities for work and recreation. Home and out-of-hours working using computer connections has some ecological justification in terms of reduction in commuting and thus of energy usage. However, there are considerable cultural risks. Home working and entertainment have already led to a widespread growth of isolationism and reduction in social contacts, which has probably impoverished the social fabric, and which cannot properly be compensated for by developing one's Internet connections. In the illusion of the "virtual" world, entertainment values become confused with life values, and both morality and ethics are easily upset.[11] One result is the growth of passivism, a docility born of having so much served up to one in fully packaged, predigested and watered-down form. It results in the growth, or at least the resurgence, of credulousness and the loss of both curiosity and critical faculty. Consent and consensus are, as Noam Chomsky maintained (Hermon & Chomsky 1988, Chomsky 1989, 1994), manufactured by manipulating the media. Independent thinking tends to be much harder work than passively accepting the diet of news, views and opinions offered by the media.

As one waits for the rise of the virtual culture, based on simulation and computer visualization, one wonders what the impact on disasters will be. Although there are undoubted research and management advantages in the more favourable depiction of catastrophic scenarios, it is easy to see the potential drawbacks. In principal, any retreat from reality is likely to be damaging.[12] The "video game" view of hazards and disasters tends to reek of "disaster pornography" and superficial attitudes to human suffering. The creation of an electronic elite will do little to ease the socio-economic disparities that are at the root of disasters. Truly, all that glisters is not gold.

Disaster and the automobile

We now turn our attention to another form of technology that has yet to receive adequate treatment by disaster researchers, despite the fact that it has been around for a century and is extremely widely used during catastrophes – the automobile.

Remarkably little attention has been given to the private automobile by students of disaster. Yet in countries where car ownership is widespread, nothing could be more fundamental. Attitudes to cars are anything but rational, and private transport has revolutionized Western society and people's relationships to each other and to the world at large. No other consumer durable has been the source of so much transformation of its image by advertising, or has assumed such a central position in economic forecasting. It is a curious paradox that the health of the automotive industry is regarded as diagnostic of the economic health of nations, yet the machine itself is a great polluter, a source of phenomenal, but avoidable, mortality, and the archetypal icon of consumer society.

---

11. I am aware that in writing this I risk being compared to those who viewed the advent of steam railways as the start of "a speedy trip to Hell". In reality all I advocate is watchfulness and a measured approach to Internet use.
12. See Mitchell (1997) for an essentially contrasting view.

Evidence is accumulating that, despite its social and economic benefits, car ownership is giving rise to a sort of low-intensity long-duration disaster in its own right (IFRCRCS 1998: ch. 2). Hence, road accident death tolls average somewhere between 500 000 and 1 million per year, and injury tolls average 15 million world wide. In Europe, where road safety has become an increasingly important issue, the number of deaths has halved since 1970, despite the doubling of volumes of traffic, although this still leaves 150 000 Europeans permanently disabled by traffic accidents each year. In contrast, volumes of traffic and accident rates are set to undergo huge increases in developing countries. For example, 4 million vehicles were on the roads of India in the early 1990s, whereas it is estimated that the figure will be 270 million by the year 2050 (IFRCRCS 1998). Road death tolls in developing nations are already disproportionately high: for instance, there are 76.8 deaths per 10 000 registered vehicles in Bangladesh, but only 1.9 per 10 000 in Japan. Hence, by 2020, road accidents may be the world's third greatest cause of premature death; they already account for more deaths than any other single factor among males between 15 and 44 years of age. Costs are also likely to soar and are already very high: it is estimated that road accidents cost the developing world US$53 billion per year. However, for small countries, accident prevention programs could have cost–benefit ratios as high as 1:10 and therefore they have good potential to reduce costs (IFRCRCS 1998: 26).

The rationale of private car ownership is to provide a source of basic transport. Yet the car is one mechanical invention in which form follows function only loosely. The result is a product that continually changes its appearance, yet with minimal improvements in functionality. Indeed, given the automotive industry's long-held opposition to innovation, the product consistently lags behind its potential, at least in terms of what could be achieved in the fields of pollution reduction, economy, safety and longevity.

In advertising terms, the result is a lie of Hitlerian proportions. The car is marketed as an object of fashion, an extension of the customer's personality, a symbol of material success and a source of liberty. How far from reality all this is! In the United Kingdom, which has a population of 57.5 million inhabitants, there are nearly 24 million cars. Evidence is rapidly accumulating that the overcrowding of roads, villages, towns, cities and parking places is causing a form of mass psychosis, the attitude of the rat in the trap. Yet it is widely predicted – on the basis of goodness knows what data – that the number of cars on the road will rise steadily to a peak of 57 million! One can only marvel at the efficiency and brilliance of advertising, which has succeeded in convincing people that it is better to fill the cities and the countryside with polluting vehicles rather than consider some rational form of alternative transport, or even do without such mobility.

Very little research has been published on the role of the car in disasters, and yet its contribution is impressive, perhaps even fundamental, in societies where there is a high level of car ownership. The matter can be divided into two inter-related parts: the car as a source of risk during disaster and as a source of reaction to the threat of impending catastrophe. Let us first consider some examples of the former case.

On Friday, 3 January 1992, fog blanketed the Autostrada del Sole, the principal trunk motorway of Italy, where it runs across the Padania Plain of the Po River Valley. In conditions of poor visibility, two large trucks collided and their flammable cargoes ignited. Some 131 other vehicles crashed into the wreckage and 33 passenger cars were destroyed by fire. Seven people were killed, 123 were injured (four of them critically) and about 250 other motorists and passengers were involved in the pile-up. In a similar event about six weeks earlier, dust storms had reduced visibility practically to zero on a Southern Californian freeway. Two articulated trucks crashed into each other and about 100 other vehicles piled into the wreckage. Both cases represent events that are so common as to merit only brief treatment in newspapers and are soon for-gotten by whoever is not involved in them.

In both cases a physical hazard (fog or airborne dust) interacted with a tech-nological system (a rapid-transit limited-access highway) to produce a risk that for some protagonists transformed itself into catastrophic loss (death and injury or massive damage to vehicles). Everything hinges on perception, but percep-tion of risk rather than perception of the physical reality. In both cases most of the damage and many of the casualties could have been avoided by slower driving, yet few of the motorists involved seem to have perceived this need and acted upon it.

In point of fact, the act of driving a vehicle involves a fair degree of escapism, a tendency that is energetically reinforced by the manufacturers and advertisers of cars. Deregulation of speed limits in the USA in the mid-1990s gave rise to about a 15 per cent rise in road casualties and an increase in the consumption of crude oil by about 50 000 barrels per day. Being predictable, these figures are unremarkable, but the lack of protest in response to them is surely extra-ordinary. Speed translates into time, and time must be saved, but in what, one wonders, is the time thus saved invested?

One corollary of escapism is the syndrome of personal invulnerability. This has been observed in various contexts with respect to the risks posed by disaster (Kilpatrick 1957), but it is particularly apposite to the use of cars. In the Big Thompson Canyon flood in Colorado in 1976, some of the motorists caught in the flash flood continued to drive until all four tyres were flat, burst by the onslaught of rocks and stones carried in the raging waters (Gruntfest 1977). Only then did they attempt to leave their vehicles and climb the canyon walls to safety – with predictably lethal results, given that the flood waters were by then at waist height and flowing strongly.[13] In Europe many of the casualties in flood disasters in mountain valleys are motorists. In such cases the car has the effect of isolating its occupants from what is going on outside, until driving is no longer possible and casualty is inevitable.

When considering the car as an agent of reaction to disaster, there is no doubt that its role in evacuation is impressive. By transforming six-lane highways into one-way streets, the authorities were able to evacuate 1.25 million people from Florida in advance of Hurricane Andrew in August 1992. In this case, wide roads, good connectivity, massive publicity and excellent coordination saved many lives. Yet it is not always so. Evacuation of the hills above Oakland, California, occurred in the teeth of the 1991 wildfires (Alexander 1995f). Amid a backdrop of skies black with smoke, families struggled to load possessions into trucks and station wagons. Bushes burst spontaneously into flame in front yards, winds howled and the winding hill roads rapidly choked with cars, evacuees and emergency vehicles. Very similar conditions prevailed in the 1983 "Ash Wednesday" bushfires in the Adelaide Hills of South Australia (Britton 1984, Voice & Gauntlett 1984). After the event one could see the eerie blotches made on the roads by cars that had been entirely melted by the fierce heat of the flames. In one instance a homeowner overloaded his car with a priceless collection of works of art from his home, only to find that the flames spared the house but burned the car and its contents to a cinder.

Even when evacuation by personal transport is completely feasible, the car is likely to prove a mixed blessing. For example, in the summer of 1987 a large landslide impounded 16 million litres of water in the River Adda valley of the Italian Alps. It was feared that the spontaneous collapse of debris barriers at the landslide toe[14] would send a flood wave several metres high through the city of Sondrio, down stream. More than 25 000 people were evacuated. But although the private car made rapid evacuation possible, it proved impossible to enforce the evacuation order and barricade all routes into Sondrio against motorists who were determined to return home. As a result, traffic jams formed on the roads across the Adda floodplain, precisely at the time when flooding was most likely and the evacuation was meant to be complete (Alexander 1988).

So many reports and theses end with the words "more research is needed" (research, after all, must be self-perpetuating if we are to sustain our own employment). But in this case it is truly so, for remarkably little is known about the role of the car in disasters. In point of fact, the role of attitudes to the car is a more appropriate research question: many owners consider their cars vainly

13. Some 136 occupants of the canyon were killed by the flood (McCain et al. 1979).
14. See Costa & Schuster (1987) for an analysis of this phenomenon.

as status symbols, investments to be protected, objects of fashion, extensions of their own personality, a source of sanctity, or a means of escaping from some of the problems of daily life. In the hands of misfits or criminals, the car can become a lethal weapon or an aid to crime. Given its role in transforming connectivity and spatial relations, the car deserves more attention from students of disaster.

Satellites and disaster

Having examined a widely diffused form of consumer technology, let us now go to the other end of the spectrum and consider a form of bespoke technology that requires very much higher rates of investment per unit produced: the satellite. Whereas the car represents tangible evidence of popular involvement in the use of technology, satellites have had a less obvious, although still profound, impact on life. For example, they cannot be seen when they are used for telecommunications, but they have nevertheless revolutionized a wide range of daily activities, such as telephoning or watching television, and have also had a profound impact on fields as diverse as logistics and international relations. Given the complexity of links between human activities and disaster impacts, it is thus unsurprising that satellites have begun to play a central role in catastrophe and its management. But it is a role that urgently requires some degree of critical evaluation, which will be offered in this section.

On 4 June 1996 the European Ariane 5 rocket, carrying four new scientific satellites valued at US$7.4 billion, exploded and disintegrated as it was launched. It was, of course, not the first spectacularly expensive failure in the space race and it serves to illustrate the high cost and considerable risk involved in such ventures (*The Economist* 1996, Hellemans 1996). Nevertheless, the pace of development remains high in all four branches of satellite research: Earth resources, meteorology, telecommunications and geodesy–geology. The hunger for data on global change has made the present day the satellite age, as the "eye in the sky" is one of the few efficient means of acquiring data of worldwide comparability.

Satellites can be useful basic instruments for environmental monitoring and data accumulation. They do not alleviate poverty, of course, and they have had only a limited impact on vulnerability to disaster. However, they are useful for monitoring and mapping certain hazards (floods, volcanic, eruptions, desertification and others), for disaster management and communications, and in forecasting events (Wadge 1994). Geostationary operational environmental

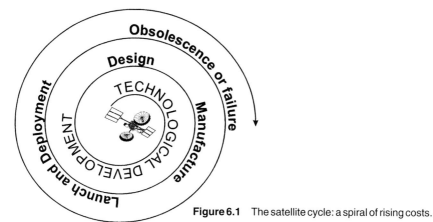

Figure 6.1   The satellite cycle: a spiral of rising costs.

satellites (GOES) are used for rapid warning of the impending arrival of tsunami waves (Bernard 1991). Digital elevation matrices can be combined with satellite imagery to create virtual-reality scenarios that help visualize landscapes and their hazards more precisely (Ehlers et al. 1989, Butler et al. 1991, Stephenson & Anderson 1997). But there are drawbacks and there has been remarkably little critical analysis of remote sensing.

To begin with, the high cost of design, manufacture and launching of satellites goes hand in hand with their often short life and rapid obsolescence. Thus, the technology requires continued commitment and an ever-increasing scale of investment for benefits that tend to be eclipsed or overtaken by new developments in technology. The result is a rising spiral of costs as governments and private consortia struggle to keep abreast of the potential of the technology (Fig. 6.1).

Given the need to justify expenditures and make them cost-effective, services and imagery are increasingly sold at high prices to potential users. With the wider availability of the products of satellite surveillance but the higher costs of their acquisition, are the skies being democratized or commercialized? As the USA has decided on a relatively open-access policy regarding the availability of space-based data, most other countries have been compelled, willingly or otherwise, to follow suit. In this sense, democracy has prevailed. But government involvement in satellite operations has often been reduced to the role of regulator, while privatization has led to apportionment of services on the basis of ability to pay (Mack 1990). Thus, disaster telecommunications do not have instant priority on communication satellites such as Intelsat, where commercial payloads tend to call the tune. In fact, protocols are needed in order to safeguard disaster telecommunications in the face of commercial priorities, especially as the former tend to be *ad hoc,* whereas the latter are mostly scheduled well in advance (Alexander 1991c).

Little critical analysis has been applied to the use of remote-sensing imagery. The cost of both acquiring and processing images tends to be high, although developments in computing have caused the latter to fall, and the former has risen in response to commercial pressures. The gradual augmentation of banks of imagery has resulted in such vast accumulations of data that one wonders whether information has not been devalued by its very copiousness. In the practice of remote-sensing image interpretation, immense efforts seem to go into ensuring that the phenomenon under study in the image can actually – or at least accurately – be detected (Townshend 1981, Southworth 1985, Congalton 1991). Moreover, to visualize a problem with better graphics is not necessarily to visualize it better conceptually. Reality is in danger of becoming too virtual and methodology too inductive, driven by the bitstream of data that have been acquired for their own sake.

The achievement of maturity in remote-sensing studies will doubtless switch the debate from what can and cannot be sensed to why it should be sensed at all. Already there are the beginnings of a change in emphasis from quantity to quality of data, and a tendency to be more selective in the use of images may yet emerge. At the heart of the problem is the modern concept of information as a formless digital commodity, rather than a source of inspiration. To redress the balance it is going to be necessary to increase the amount of critical analysis of what remote sensing *actually* adds to the understanding of environmental problems (Belward et al. 1994). The current challenge is not merely one of integrating satellite data with other aspects of disaster mitigation and management – and other technologies used to this end – but of doing so in a way that maximizes real benefit while minimizing expenditures.

The notion that environmental remote sensing is a cost-effective way of tackling environmental problems led to the setting up of SPARRSO, an image receiving and elaborating centre in Dhaka, Bangladesh. In August 1988 SPARRSO was inundated and forced out of action by the very floods it was seeking to monitor (Rasid & Pramanik 1990). This curious incident leads one to reflect on the question of equity in access to information. Although such countries as India and China have their own satellite programs, Europe, Russia and the USA have dominated the use of near space. Satellite images and facilities have been "donated" to Third World Countries as part of development aid and disaster relief, but, like all such donations, the inputs have been patchy and intermittent. Given that some of the greatest potential for the use of satellites is found where other sources of communication, monitoring or mapping are meagre, the world community needs to give more thought to this aspect of technology transfer (Hassan & Luscombe 1991). The danger is that cost considerations will inhibit it, and also that the results will be poorly tailored to local needs – "remote" sensing in more ways than one.

The world's hunger for digital data apparently cannot be abated. There is a risk that problems will be appreciated or viewed as solvable only if they can be converted to bytes and to pixels on a screen. What is important is to ensure that the acquisition of a solid base of environmental data, and robust networks of telecommunications, be viewed not as an end in itself but as the beginning of a process of mitigation that deeply involves questions of ideology and equity.

By increasing the flow of information, reducing communication times and making visible what is on the ground, satellites have contributed to the cohesion and democratization of the world. But the 1991 Gulf War showed how easily the Information Age can be converted into a Disinformation Age by the adroit manipulation and suppression of data (Greenberg & Gantz 1993).[15] There is an important lesson here that warns us not to take too much on trust. Rising expenditures on technology, and its increasing use in disaster management, prompt a few words on the current status of catastrophe mitigation. This is especially pertinent, given that technological advancement has been both a cause of rising losses and a means of counteracting them.

Mitigation and the rising toll of losses

At the dawn of the twenty-first century, great strides have been made in the field of disaster mitigation. The problems posed by natural and technological catastrophe are now well known and a substantial body of knowledge has been assembled on how to tackle them.[16] Although the field of disaster studies is young, it has assembled an impressive body of theory. Yet, much of the existing theory of disasters is ill equipped to explain the salient fact that costs and losses continue to rise, and at an exponential rate (Berz 1994). In 1989 the Loma Prieta (California) earthquake cost an estimated US$12 billion (Tubessing & Mileti 1994). In 1992 Hurricane Andrew caused US$18 billion in losses in Florida (Rapaport 1994) and it was feared that the next large disaster could cost as much as US$40 billion (US National Research Council 1992). Then in January 1995 the Kobe earthquake in Japan gave rise to costs and losses valued at US$131.5 billion (Japanese IDNDR Committee 1995). In July 1996 Hurricane Fran caused US$3.5 billion in damages in South Carolina alone. And yet the potential for major

15. Allied claims of high precision in bombing missions were belied by later reports that attributed about 10 per cent accuracy to the bombs, but that information was released long after public interest in the war had waned.
16. See the 12 volumes published in the 1970s and 1980s by the UN Office of the Disaster Relief Coordinator and entitled *Disaster prevention and mitigation: a compendium of current knowledge* (e.g. UNDRO 1984).

disaster losses in the future is considered to be even higher.[17] At the same time, death tolls, always somewhat variable, have remained fairly static in aggregate (IFRCRCS 1997: 117). It is difficult to find a yardstick against which to judge trends in mortality. Disease rates are one potential standard, as the probability of death by disease tends to form a limit to the level of tolerable risk in other fields (Kasperson & Kasperson 1983). Advances in medical science have increased life expectancy in this regard, but, for a substantial proportion of the world's population, curable diseases are currently resurgent as a result of poverty and lack of hygiene.[18]

Nevertheless, if we consider that the population-doubling time is in the order of 19–24 years in some African countries, to keep death tolls constant, risk levels must in effect halve during this time, a very difficult objective indeed. Hence, the relatively slow increase in death tolls is testimony to mitigation and relief efforts. But the counterpart is the steep rise in the cost of damage. In part this must reflect the impact of inflation and the constant revaluation of property (Berz 1994). But it must also indicate a failure to stem the growth of vulnerable capital goods in the zones of major risk. Moreover, life is becoming ever more complicated, and hence the cost of repair and replacement and the demands of security require ever higher levels of investment. As the trend indicates, this is both a worrying and an unsustainable situation. The escalation of costs has alarmed the re-insurance industry (Berz 1991, 1992, 1994, Royal Society 1998), which fears for the survival of insurance companies under the increasing burden of claims. And although death tolls have not risen in the same way as damage costs have, the increasing vulnerability of capital goods bodes ill for public safety.

The accumulated body of theory on hazards and disasters is perfectly able to explain the sources and significance of human vulnerability, but it does not account adequately for the rising tide of losses. In my view this is because it is poorly linked to the momentous changes occurring in society at the turn of the millennium. When the costs are so high, it is inevitable that disasters become big business. They are, moreover, distinguished by sudden and arbitrary transfer of capital. The processes that govern these transfers are those that shape society as a whole. Hence, although disasters are traditionally viewed in terms of welfare (Leitko et al. 1980), in reality they have much to do with the logic and organization of market economies, global commerce (and its geopolitical strategies) and unfettered capitalism. This leads us to a more detailed consideration of the economics of disasters in the modern world.[19]

---

17. The final cost of the 1994 Northridge earthquake in California is estimated to be US$42 billion.
18. See WHO (1986) and Aghababian & Teuscher (1992) for an indication of how communicable diseases relate to disasters.

Economic growth and disasters

The unification of the global economic system, which has been under way since early colonial times, has been virtually completed by its ability to move vast sums of money around the world more or less instantaneously using modern telecommunications. The mobility of capital has, at least temporarily, outwitted labour, which is much less mobile (see Fig. 4.8).

There is some uncertainty regarding the extent to which disaster mitigation depends on economic circumstances. On the one hand, simple measures, low-level technology and good social organization can increase human safety appreciably. But on the other hand, good mitigation requires sustained levels of monetary investment, whereas poverty induces marginalization and is highly correlated with vulnerability to disasters. Moreover, economic instability is not conducive to good mitigation, and in many cases the transfer of technology depends on the availability of sufficient funds. This should lead to a debate on the meaning and nature of world economic growth in the context of disaster losses. For more than 20 years the value and necessity of growth have been questioned (Meadows et al. 1972, Pestel et al. 1989), but nevertheless, the conventional model, that the world must produce and enrich itself materially, remains firmly entrenched at the heart of economic policy.

In 1996 the United Nations *Human development report* revealed that the collective wealth of about 360 billionaires exceeded the combined annual income of countries that contain 45 per cent of the world's population (Solow 1996). Comparing the economic condition of the richest fifth and the poorest fifth of the population indicates that income differentials have more than doubled in 36 years. As 89 countries are worse off than they were 15 years ago, the report concluded that economic growth has failed for one quarter of the world's population. Moreover, two fifths of countries with overall economic growth have not seen this reflected in growth in employment rates. The report classifies it into five kinds, growth which:

- does not translate into employment (jobless)
- is not matched by the spread of democracy (voiceless)
- extinguishes cultural identity (rootless)
- despoils the environment (unsustainable, futureless)
- and in which the benefits are expropriated by the rich (ruthless).

---

19. Important works on this subject include the following: Dacy & Kunreuther (1969), Russell (1970), Douty (1977), Natsios (1991), US National Research Council (1992), Britton & Oliver (1997) and Kunreuther & Roth (1998).

Disaster mitigation is likely to be at variance with all of these kinds of economic growth, for it cannot succeed on the basis of excluding people or their environments from the benefits of protection (UNDP 1996: 2–4).

In many respects the world economy is both founded on and motivated by inequality. Consider some of its better-known problems, which can be grouped into three categories: the straitjacket of international trade, the pressures of militarization, and the problems of debt. In a poor country, growth may be seen to depend on the exploitation of primary resources that are largely manufactured and consumed abroad, but commodity prices may be controlled or maintained at artificially low values by the dominant purchasers. Multinational involvement in resource exploitation (logging, mining, plantations, and so on) may reduce the opportunities for government control of such activities, if the government is willing to control them in any case. The result may be both an increase in environmental insecurity and greater physical vulnerability to hazards such as floods and landslides (Albala-Bertrand 1993). Emphasis placed on exports over domestic consumption in order to generate foreign exchange may lead to impoverishment and increased marginalization and vulnerability. One consequence is the emergence of domestic black markets and parallel economies, by which people endeavour to survive in the absence of official support. These are difficult to deal with after disaster because they complicate attempts to restore healthy trade by making sensible use of relief commodities.[20]

As noted in previous sections, international aid is often granted on the basis of strategic alliances rather than demonstrated needs. One aspect of this, which has by no means disappeared after the Cold War, is militarization. As the industrialized world reduces the size of its armies and moves towards greater technological investment in weaponry, it continues to exert pressure on the developing countries to arm themselves, a pressure exacerbated by regional tensions and rivalries. For such countries, armaments are obviously unproductive, both in terms of generating economic security and with respect to disaster mitigation. They are bought at the price of diverting scarce civil resources from such goals (COPAT 1981, Pugh 1998).

Over the years, public-sector borrowing in poor countries has been fuelled not only by military considerations but by the allure of large-scale development projects, often with the active encouragement of developed nations that supply the capital, the technology and the contractors for such projects. Economic downturns hit such countries very hard, and debt may become an uncontrollable problem. Structural adjustment programs and rescheduling arrangements impose strict fiscal criteria for the repayment, reduction or containment

---

20. Perhaps because it is a somewhat elusive phenomenon, there has been little analysis of the role of black markets and parallel economies in the context of disaster.

of debt and the future limitation of public-sector borrowing. But, even when they succeed, such financial programs seldom benefit the poor, who nevertheless have to shoulder the burden of harsh austerity measures. In this way, development is achieved at a high price: the benefits to local people of large development projects are dubious, the risks are high and the costs may be unsupportable. Increased environmental risks are combined with greater poverty, itself an invitation to disaster (Blaikie et al. 1994).

This observation brings to mind the plight of Central America in the wake of Hurricane Mitch, which moved slowly and devastatingly across Honduras, Nicaragua and neighbouring countries from 26 October until 3 November 1998. Death tolls were estimated to be 17 650, including 12 000 in Honduras and 3000 in Nicaragua, where more than one third of the fatalities resulted from a single lahar on the flanks of Casitas volcano. Of the 4.3 million people in 7 countries who were affected by the disaster, 2.7 million were in Honduras (representing 41 per cent of the population) and 800 000 were in Nicaragua (16 per cent of the population). In Nicaragua, 70 per cent of roads were damaged and five departmental capitals were cut off from road access. Of agricultural production, 70 per cent was lost in Honduras and a similar proportion was devastated in Nicaragua. It was estimated that the disaster has set back development processes by 20 to 50 years.

The developed world's initial response was extremely cautious: the affected countries were to be assisted to the tune of one tenth of the sum voted at about the same time by the Group of Seven to sustain a single hedge fund that had run into difficulty through excessive speculation and required $2 billion to keep it afloat. Cynics would find much to confirm their attitudes in the nature of foreign involvement in Central America. For example, the Tegucigalpa golf course (mainly used by expatriates) was still being watered in the days after the hurricane, although it faced a slope on the other side of the valley where houses had been reduced to rubble by slumping or washed away by swollen streams. More fundamentally, the Chiquita banana corporation, whose finances dwarf those of the Honduran government, dismissed most of its 40 000 workers in Honduras in order to minimize the economic impact of the hurricane on its operations. There was little sign that the rich multinational tropical fruit producers would step into the welfare gap that national governments and governmental aid were so signally failing to fill.

The key to the problem appears to be the direct relief of current poverty. One economic expert has argued that the present fashion for sustainable development has diverted attention from this simple but intractable goal (Solow 1996). The lack of a global political will to tackle the problem is tragic: a 0.25 per cent tax on international financial transactions with the proceeds recycled into poverty relief would achieve lasting benefits, but there is no sign that it is under serious consideration by world leaders.

In 1990 the United Nations instituted the International Decade for National Disaster Reduction with the aim of reducing the roll of disasters by half during the decade 1990–2000. It should be remembered that during the previous quarter of a century the numbers of people affected by disaster had risen steadily to a mean of over 200 million per year (IFRCRCS 1998). The IDNDR secretariats and committees have worked hard to encourage mitigation of natural catastrophe, but the problem remains as serious as it was in 1990. What is needed, and what is lacking, is a solid alliance between global disaster reduction initiatives and worldwide efforts to alleviate poverty, marginalization and vulnerability. Some 1120 million people live below the poverty line, in the sense that they cannot afford a minimum diet that is nutritionally adequate and they lack essential non-food items such as medicines. The homeless number 100 million (Wisner 1998). For such people, disaster mitigation has relatively little meaning while life in general is so precarious. UN agency spending on disaster relief reached US$1875 million in 1993, and Red Cross and Red Crescent spending was US$792 million (IFRCRCS 1996). But high though these figures are, they are tiny by comparison with both mitigation needs and world capital flows. Much more financial commitment is needed if the disaster problem is to be solved.

The changing face of emergency management

Let us now turn from the global view to a more local approach and consider how emergency management has changed in recent times. This will bring us back to the initial theme of this chapter: the relationship between society and technology with regard to disasters.

As Thomas E. Drabek and George M. Hoetmer put it, "The role of the emergency management professional is neither easily performed nor well understood. Nor is it a readily accepted role in many local jurisdictions" (Drabek & Hoetmer 1991). But despite this there has been a process of evolution in the standards and techniques of emergency management: progress has been made towards greater security and professionalism. Let us review that progress and examine some of the problems and dilemmas that stand in the way of better disaster management.

Perhaps 90 per cent of emergencies do not require special procedures and organizations to deal with them (Drabek 1986). For the rest we may suppose that there is a qualitative change in the exigencies and requirements of management. This begs the question, also considered elsewhere in this volume, of just what a disaster *is* and how it differs from a "routine" emergency?[21] Various efforts have been made to define the phenomenon quantitatively as an index based on length of forewarning, and scope, magnitude and duration of impact (Foster 1976, Burton et al. 1978, 1993, Johnson 1990). There have also been attempts to define the severity of a disaster according to its impact ratio: the volume of losses to remaining resources (Tobin & Montz 1997: 23). But it has also been shown that context is of paramount importance: if the public perceives a series of incidents to represent a betrayal of security measures, then it may tend to exaggerate their significance (Horlick-Jones 1995a). In synthesis, the context of risk and mitigation can turn a major incident into a minor disaster, at least in terms of how the public views it (Mitchell et al. 1989).

It is axiomatic that no standard blueprint exists for an emergency management structure. However, in principle, such a structure should be compatible with the procedures and organizations used to manage "routine" emergencies: it should extend them, not supplant them. The first principle here is that emergency planning and emergency management are linked activities: the one is indispensable to the other. Disaster planning is an essential prerequisite for the creation of an emergency management structure (it cannot be improvised successfully in the heat of the crisis; Foster 1980, Lindell & Perry 1992). A good plan should not be too detailed, as this would restrict opportunities to improvise creatively and to maintain flexibility. But it should provide for all the main needs in the emergency: the care of survivors and the injured, the mobilization of emergency personnel and resources, the management of emergency activities, the restoration of essential public services, the assessment of damage, and efforts to guarantee public safety, warn the public and keep it informed of developments, plan the recovery and keep adequate records. Good emergency planning is based on a combination of prior preparedness and *ad hoc* improvisation. The former should be continuous, both educating the participants and building upon their stocks of knowledge (Daines 1991). However, experience suggests that government organizations at various levels often resist emergency planning (Davis & Seitz 1982): education thus involves demonstrating the need for improved preparedness. Paradoxically, if such efforts are highly successful, a well managed disaster is less likely to stimulate demands for better mitigation than a badly managed one.

The process of emergency management has come a long way since the 1950s, when it largely represented civil defence, an outgrowth of military activities (Fritz & Marks 1954, McGill 1957). Since the height of the Cold War there has been a change to a professional all-hazards management approach that has recast the role of the emergency manager (Drabek 1991). The political profile of

---

21. E. L. Quarantelli has tackled this problem eloquently on various occasions (Quarantelli 1978, 1989, 1995, 1998).

the job is now considerably stronger, and its incumbents must have plenty of political acumen and ever greater administrative and communications skills. The qualities of a good emergency manager include serendipity, intuition and level-headedness. Above all, he or she must be able to make sense rapidly of large quantities of conflicting information. Successful managers are not autocrats but are compromisers, facilitators, mediators and integrators. Their sense of vision must not exceed their grasp of the situation. Hence, they must be realists, for they may have to cope with high levels of risk, fluid and ill defined situations, competition and conflict among emergency workers, and urgent situations in which time is scarce (Drabek 1990). Moreover, emergency managers must live with the enduring legal, political and social consequences of decisions that they take during the emergency. This is rendered all the more complex by the multiplicity of roles that the emergency manager must assume; for example, administrator, planner, organizer, staff supervisor, leader and arbiter of disputes. Many of these roles must be created before they can be exercised, which is part of the emergency-planning process. Moreover, programmed and *ad hoc* decision making often require different skills, which casts the emergency manager in a dual role (Drabek 1987).

Over the years there has been a gradual change in the basic model of emergency management. Traditionally, it has been based on a chain of command and a rigid hierarchical structure of authority (Foster 1980). This has the advantage of being a robust system in which roles and responsibilities are clearly defined. Centralization ensures that decision making is not duplicated and information is collected and evaluated where it can best be used. But this model has been subject to increasing criticism for being too rigid and mechanistic, and for allowing bottlenecks to occur in the flow of both decisions and information.

The alternative model, which has become very fashionable, has acquired a series of different names: incident command system (ICS; Irwin 1989, Sylves 1991), emergency-resources coordination model, matrix organizational structure, multiple-agency coordinating system, integrated emergency-management system, and so on (Fig. 6.2). They are all variants on a theme in which adaptability is the key criterion. A good ICS must be expandable from the scale of minor emergencies to that of full-scale disasters. It should be based on the procedures used by existing agencies and must be simple enough to be inexpensive and easy to learn. It must be adaptable to new technologies as these are invented and deployed, and it must be capable of handling a wide variety of different types of emergency, including secondary and multiple hazard impacts (Denis 1997). Lastly, it must function both for single and multiple organizations and for single and multiple jurisdictions.

A three-stage procedure is used when setting up this kind of system. First, all hazards of any significance in the region are assessed. Secondly, an inventory

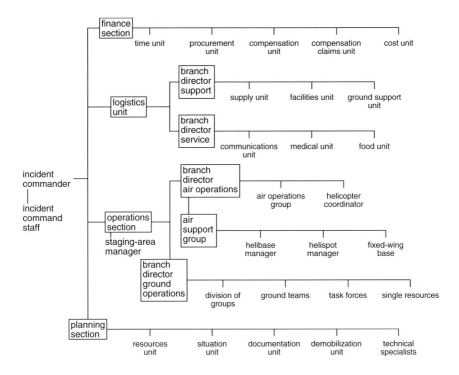

Figure 6.2 Structure of an incident command system

of resources and capabilities is compiled and, thirdly, the system is designed in order to reduce the gap between vulnerability and capability (Irwin 1989).

Under ICS rules, the chain-of-command is not necessarily rigidly adhered to. Rather than "command and control", emphasis is placed on coordination through consultation and flexibility. Job assignments are not rigidly defined, as task forces can be constituted to deal with problems as they arise. The important thing is to arrive at a consensus on the goals to be achieved and to move towards them by a process of delegation, participation and mutual involvement. This requires that all organizations participating in the emergency have common elements of procedure, terminology and structure, in order to sustain the flow of information that makes the collaborative mode possible (Belardo et al. 1983, Cate 1995, Fischer 1998).

As stated above, the principal exigencies of the emergency period will be a shortage of time, with pressure to take action quickly, limited and uncertain flows of information, shifting priorities (known as "agent-generated" and "response-generated demands", according to whether the imperative comes from the hazard itself or the damaged social fabric; Kreps 1983), and the need to reconcile overlapping lines of authority and responsibility. The key to a workable ICS is to have a functioning emergency operations centre (EOC; Scanlon 1994). Obviously it must be situated where it is most useful, without falling prey to the disaster itself, and must be able to sustain robust lines of communication under circumstances of great duress. Its tasks will include ensuring that personnel and resources are mobilized efficiently and without unnecessary duplication. The public and news media must be kept informed of developments, and there must be effective liaison with regional and national governments in order to bring relief resources into the disaster area. Above all, the EOC must maintain open communication between its participating organizations, as information flow is the key to the collaborative mode of emergency management (Irwin 1989).

Given the tendency for emergency management to avail itself of increasingly high technology, the setting up of an emergency operations centre represents a considerable investment, and one, moreover, that creates a structure destined not to be used for the majority of the time. For this reason, there may be considerable political opposition to the idea, although when disaster strikes it will soon become apparent how difficult it is to do without an EOC. Clearly, it is easiest to justify the investment in particularly hazardous areas where emergencies are relatively commonplace and the EOC is unlikely to gather dust. Where disaster is rare, the costs can be reduced somewhat by relying heavily on volunteer groups (which must be specially trained, organized and well integrated into the emergency management structure; Moran et al. 1992) and using an "embedded" rather than an independent structure. In this, the civil protection workers are drawn from existing organizations and they simply change their roles during the emergency; full-time professional emergency managers are not used. This has the drawback of requiring workers (such as fire chiefs and technical officers) to perform radically different tasks under emergency situations in comparison to how they operate in normal work situations.

Despite the need for order and prior planning, improvisation is also a vital part of emergency management. **Emergent groups** such as relief committees will be formed for the occasion; **emergent norms** will develop, representing new modes of collective behaviour and **emergent social structure** will define new relationships, networks, hierarchies and divisions of labour (Perry et al. 1974, Saunders & Kreps 1987). All this presupposes a strong desire to work collectively (which the emergency manager must learn to exploit for the good of the community), but it is also true that the altruism inspired by disasters can dissipate rapidly. The emergency manager may well have to face challenges to his authority (which amount to equivocation in the chain of command, whichever method of disaster management is adopted), and will have to bear the legal risks of human error, incompetence or substandard performance (Kusler 1985, Schneider 1992). At the same time the public may well choose to do as it pleases and to ignore binding instructions from public officials. In the extreme cases that are characteristic of the individualistic culture of parts of the USA, householders may mount an armed defence of their properties against presumed looters, which can render emergency work even more hazardous. Emergency respond-ers may suffer from conflict of roles between their responsibilities to their families and to their jobs (although this can be avoided by planning to take care of key officials' family members; Foster 1980). All emergency workers may be liable to critical incident stress (a form of post-traumatic stress disorder), and the principal officials may end up suffering from the Jehovah or *magna mater* complexes (Duckworth 1986; Alexander 1993: 568–9), which involve a tendency to assume too much responsibility. In short, no amount of prior planning can prevent emergencies from being messy, unpredictable events.

Since the 1950s there has been a considerable growth in social research on disasters, especially regarding the structure and dynamics of organizations under the stress of emergency situations.[22] As always, there has been a lag in transforming research results into positive changes in the way that emergencies are managed. However, it is now clear that social research *has* had an impact on emergency management procedures, at least in the industrialized countries (Auf Der Heide 1989, Drabek & Hoetmer 1991). But there is a remarkable hiatus between the position of academic researchers and that of disaster managers. The latter are for the most part not high-flying social scientists but are essentially pragmatists. On the other hand, academics seem to revel in obfuscation. Hence, the language of social research is replete with jargon – "social construction", "functionalism", "claims-pressing", "emergence", "agent-generated", "system-atic", and so on.[23] But the opacity of meaning that seems to confer legitimacy on academic publications also imposes a barrier between the generators and the users of knowledge. In this respect it is as well to remember that the roots of "disasterology" involve the marrying of the theoretical and practical: if it will not work in the field, it will not work in theory either.

Discussion of how disasters are managed points to the need for an example. Let us combine the instructive elements of actual cases into a hypothetical scenario that may serve to illustrate the problems of emergency management and expose some of the weaknesses in how it is carried out.

---

22. Drabek (1986) has provided a classified guide to social research up to the mid-1990s. See also Dynes & Tierney (1994).
23. For example, see Ploughman (1995), an otherwise very useful article. One should be thank-ful that the field has not been touched by the post-modernist debate, with its positively baroque jargon and highly elusive meanings.

A scenario

Scenario modelling is widely used as a means of exploring the possible consequences of certain events and actions. It involves setting up a chain of events and developing the sequence of outcomes by means of a progression of explanations ordered over time (Foster 1980, Borchardt 1991). To be effective as an explanatory tool, the scenario must describe events and outcomes that are plausible, likely, significant and logically consistent. Given that most disasters involve an element of chance, the more complex forms of scenario modelling can evaluate more than one possible outcome. The methodology allows its developer to consider the situation and draws conclusions from it, not only regarding its final outcome but also in terms of any point or stage during the evolving crisis and its aftermath. Here we use a simple version of the method and draw conclusions only from its final outcome.

Consider the following rather depressing scenario, which illustrates the connection between natural and technological hazards and between rural and urban development. It is based on an amalgam of several real examples.

One Sunday evening an earthquake severely damages a wide area of a southern European country. This includes an industrial growth pole, which is situated in a rural valley among low hills and is connected to the nearest urban centre by a modern limited-access highway. The site includes a pharmaceutical factory that produces toxic chemicals by a continuous high-pressure process. Damage to the reaction chamber and control valves causes an extreme chain reaction that raises the pressure to several atmospheres. This is sufficient to rupture the containment vessel, which was not designed to resist horizontal acceleration and had begun to shear off its stays during the tremors. The resulting explosion blasts a hole in the external wall of the factory and vents about 40 kg of highly toxic chemicals, which are light enough to remain suspended in the air. The toxic cloud begins to drift northwestwards at about 5 km/hr towards the largest town in the region, which is situated about 10 km away. It is a hazy evening and the cloud passes undetected.

At the same time, in an adjacent factory, a concrete and steel tank containing about 1000 litres of concentrated sulphuric acid is toppled by seismic acceleration. As it crashes to the ground, it splits open and discharges its contents rapidly into an adjacent river, where a water discharge of about $3\,m^3$/sec carries them at a velocity of 1.3 m/sec down stream, southeastwards, towards another large town. If it does not dilute sufficiently in the flow, the acid will threaten to infiltrate shallow groundwater aquifers and accumulate in a small storage reservoir from which the town obtains its water supplies.

As it is Sunday evening, practically the only personnel at the industrial site are a few security guards and nightwatchmen who have not been fully trained to deal with industrial accidents. They have no respiratory gear and cannot locate any such equipment stored in the factories. Electricity supplies have failed as a result of damage to a nearby substation, and auxiliary supplies, which are provided by battery, are insufficient to reveal the full scope and nature of the damage.

A cursory attempt is made to assess the situation, which results in one night-watchman being overcome by toxic fumes. He is carried to the relative safety of a factory gatehouse. From inside, colleagues attempt to telephone the police and emergency services. However, a combination of damage to the local telephone exchanges and overloading by members of the public who are frantically trying to call relatives, friends and the emergency services, makes it impossible to get through.

One security guard is despatched by car to the nearest town, but he will have to negotiate a series of hazards. To begin with, in the darkness (there is very little moonlight) it is impossible to know whether the highway, sections of which are elevated on concrete pylons, has been damaged to the extent of making it precarious or impassable. Secondly, where it passes across the river, concrete bridge abutments may have been corroded by acid. Thirdly, landslides and rockfalls may have obstructed the carriageway. Fourthly, further damage may occur at any moment during aftershocks. Finally, if the local emergency services are functioning, they are likely to be engaged fully in dealing with casualties and rescue operations in the towns they serve.

Disaster planning in the local area has been cursory. Although they are listed as producers of hazardous materials, the industrial plants have not fully obeyed the European Union's Seveso II Directive on mitigating industrial hazards (Wettig & Porter 1997), as interpreted in national legislation that has not been strictly enforced. In the interests of encouraging regional development, government control in this respect has been lax.

The personnel on site have no alternative means of telecommunication and no adequate disaster plan. Besides telephoning some key officials, such as the plant directors (which proves to be impossible, given the problems with the telephone network), or attempting to visit their houses (which is logistically difficult), the nightwatchmen and security guards are unsure of where there are experts who have knowledge of industrial incidents and how to contain them. At this time two dangerous vectors are travelling in opposite directions through the night: the toxic cloud is heading unannounced for one urban area, while the acid-contaminated streamwater is travelling unheeded towards another and may soon enter its water supply (which has not been rendered inoperative by the earthquake). In the former case, people are made more vulnerable by remaining out of doors because their houses have been damaged, because they are afraid to go back into them, or because they are helping with search-and-rescue efforts. In the latter case, there is some chance that people will remember not to drink tap water until its safety has been verified. If the acid does not dilute sufficiently before it arrives in the nearest urban area, it may do considerable damage to pipes and culverts, as well as to the local water-treatment plant.

In the end, crosswinds disperse the toxic cloud with only light non-fatal casualties among the rural population, while the acid dilutes and is neutralized by naturally occurring alkalis to the point that it is no longer a serious hazard. The highway turns out to be largely undamaged and national government funds are rapidly disbursed in order to rebuild the factories. There is much angry discussion among the public and in the news media when the full story comes to light, but, quite simply, priorities lie elsewhere in problems such as unemployment, unaffordable and damaged housing, and lack of regional development. Thus, most of the mitigation measures hastily promised in order to placate public opinion are not adopted. Neither is disaster planning greatly improved: there is a sense that, now the disaster has occurred, preparations for the next one can wait until times are easier and the region has recovered from the economic blow dealt by the current damage and destruction. Above all, through lack of political will, plans to contain industrial accidents are not integrated with municipal and regional disaster plans. Thus, the next time disaster strikes, the results may be considerably less fortuitous.

In this scenario the context appears to be more important than the event.[24] The environment is one of a depressed region with a fragile tenor of economic development. There is a strong, historically rooted sense of fatalism with regard to disasters and it is not helped by the cultural and scientific isolation exacerbated by lack of integration between the region's institutions and the rest of the continent. There is a lack of funds, expertise, political will and public pressure for better mitigation of disasters. As a result, there is a chronic, unfulfilled need for hazard microzonation and response mapping, which should be based on a modern all-hazards approach. Interactions among hazards need to be studied at the local level. Disaster plans need to be compiled, integrated, coordinated and harmonized with each other for factories, companies and corporations, emergency services, municipal governments and regional administrations. Campaigns of public education about hazards and disasters are required, and support needs to be generated for mitigation efforts (Dudley 1988, Maskrey 1989).

The scenario illustrates how the post-disaster "window of opportunity" for mitigation is neutralized by context. Blame and recrimination are strong forces, but they are short-lived, whereas risk is enduring. But disaster tends to be considered as one of the more normal unexceptional risks of life, given the way in which it is constantly overshadowed by the problems associated with poor

---

24. See Mitchell et al. (1989) for a comparable example.

economic performance and low average incomes. Moreover, history has periodically visited disaster on the local populations to the extent that its coded message is one of acceptance, scarcely counteracted by modern positivism, the "can-do-ism" of the technological age. A political culture of grace and favour does not foster major initiatives unless they can be integrated into existing social and organizational structures. Many appointments are made on the basis of patronage, not competence: in a new field such as disaster prevention and management it is difficult to import and fully utilize the necessary expertise. Both public perception and political expediency favour spending public funds on reparation (which becomes an unavoidable commitment after disaster has struck) over disbursing money on future mitigation, even though cost–benefit ratios (which are never calculated) are likely to be weighted heavily in favour of the latter.

Under the circumstances, the stimulus for better mitigation must come from sources other than the direct impact of disaster. The winds of change blow slowly. A key role is played by national legislation on disaster preparedness, mitigation and planning. Regional and local government must be both compelled by national government to enforce it (the stick) and given the necessary financial resources to do so (the carrot). Paradoxically, this is hindered both by devolution, which allows too much autonomy at the regional level, and by centralism, which makes regional initiatives the victims of tax cuts and budget shrinkage.

This is perhaps typical of certain situations in developing regions and developing countries. It is one example of how the First and Third Worlds often have to face the same problems.

CHAPTER 7

Moral and philosophical issues

This chapter shifts the balance of argument from technology to moral, philosophical and social issues. In the modern world, technology is, of course, a major influence on these. Hence, what we often encounter is a subtle, inadvertent mutation of the time-honoured existential questions, as the methods and mechanics of technological change impinge insidiously upon the relationship between people, their perceptions, and environment in disaster. We begin with warfare, a corollary of natural disaster and at least as pervasive a problem in the modern world.

Natural disasters and armed conflict

Pliny the Elder recorded that a large earthquake occurred during the second Punic War in 264 BC, at the height of the Battle of Trasimeno. The ground shook vigorously, towns were ruined, landslides cascaded down hillsides, and rivers were impounded or rerouted. But the fighting did not stop. Hence, the carnage and destruction caused by nature were added to those produced by mankind's bellicosity (Alexander 1984). Yet, although this one example describes a mere coincidence, it highlights the need to view the two faces of the Janus of destruction – wanton and inadvertent – in the context of each other. In the modern world, armed conflict has become such a pervasive influence on life that, in many of the countries most afflicted by natural disasters, it provides a constant setting; warfare is disaster carried on by other means.[1]

To begin with, there are obviously many forms of warfare. Conflicts may be formally declared, informally prosecuted, or carried on by clandestine means. International wars, involving invasion and cross-border bombardment, are complemented by civil conflicts, but given the frequent propensity of the latter

---

1. See Macrae & Zwi (1992) for details.

to stimulate outside intervention – the two are not mutually exclusive. These are usually forms of intensive conflict, but there is also a tendency for low-intensity conflict to proliferate in the form of guerrilla activity, insurgency, terrorism and torture. Thus, the spectrum of conflict ranges from mere armed tension and sporadic incidents to all-out warfare. Yet, such is the destructive power of modern weaponry, and its copiousness in the field, that the two ends of the spectrum have begun to approach each other, for destruction can be achieved on a large scale with greater efficiency and less exertion than ever before.

Two great conflicts marred the twentieth century, but since the second of these more than 150 local and regional wars have broken out. The early part of the century saw warfare globalized and the latter part has seen that tendency continued by superpower polarization and the widespread revivals of nationalism, separatism and fundamentalism, not to mention the persistence of the age-old desire to dominate and subjugate. Immense investments in technology and strenuous efforts to market it have made warfare a vastly more efficient destructive process than ever before. This leads to some oddly ironic contrasts. A glance through the international catalogue of armaments, *Jane's weapons systems*, reveals that it is full of advertisements for cannonry that will pierce the best available defences and armour that will resist the most powerful cannons. One cannot have it both ways. The sheer intensity of technological, economic, social and political investment in warfare means not only that wars often persist for decades without conclusion but that *total warfare* is a much more common phenomenon than ever before. Some 90 per cent of the victims of war in these past 50 years have been civilians, not soldiers, and a majority have been women and children. At the same time, the global threat of nuclear war has extended the potential front line of conflict to every citizen of our planet (Bunge 1966: 277; Sagan & Turco 1991), and has induced both brinkmanship in international relations and precisely the apprehension that this would be expected to produce.

In peace or in conflict, warfare is less remote than we may think, for, even in countries at peace, the global economy is strongly based on the so-called military–industrial complexes that design and produce armaments. Defence, as it is euphemistically known (its borderline with aggression is diffuse and permeable) is a major expenditure. Moreover, conflict has resulted in over 50 million refugees, and only the displacees have been counted, not the ones whose lives have been ruined or disrupted but who have no status under the Geneva Convention of 1951 (Loescher & Monahan 1990, Toole & Waldman 1991).

Although rape and pillage are doubtless as old as armed conflict itself, the traditional view of military action is still one of clearly defined enemies who attack each other in well planned battles. But modern wars have begun to assume the guise of an all-out assault on the civilian population and its environment.[2] From the napalming of Vietnamese tropical forests to the "ethnic cleansing" (an obscene euphemism) of Yugoslav peoples, the strategic objective has shifted subtly from denying the enemy cover to total eradication of the opposition. The new weapons not only include mortars, grenades and tanks, they scorch the very ground of the theatre of conflict. Millions of landmines have been sown, and most often where they will do maximum damage to civilian women, children and livestock. Rural people are terrorized and tortured, women are raped, and children drugged and conscripted (they tend to make compliant, obedient soldiers). Property is expropriated and wantonly destroyed, vegetation is defoliated or burnt down, resources are hoarded or impounded. Aid is denied to civilian populations in order to make them compliant, food is rationed in order to weaken them, and if they do not bend to the will of the military, slavery and conscription may await them (Macrae & Zwi 1992).

Let us now compare some of the negative characteristics of natural disaster with those of warfare. When drought, flood or wildfire occur against a background of military conflict, the repeated impact of extreme geophysical events may lead to progressive environmental degradation (Varley 1993, Black 1994). Weakened by the dual attack of man and nature, social systems may fail to recover adequately between one disaster and the next and may fail to develop sustainable coping mechanisms. Thus, lack of protection of civilian populations leads to social and personal insecurity.[3] The territoriality of war makes access to land difficult and tends to reduce people's means of coping. Strongly deteriorating conditions may be accompanied by atrocities. The logical conclusion is that, where they occur together, natural disasters and armed conflict are weakening influences on populations.

Warfare is often a cause of weakened environmental resistance to disaster. To begin with, defoliation by bombing or napalming slopes in humid tropical areas can make them especially unstable in times of seasonal rain and can vastly increase landsliding, water yield and the sediment load of streams. Conversely, fire has for centuries been a chosen weapon or inadvertent consequence of fighting and it can of course be immensely destructive to ecosystems when used more intensively or more often than nature would do, especially if it becomes more frequent or more intense than to permit recovery of the vegetation and soil. In this context, warfare has traditionally included an element of environmental vandalism, for example by breaching sea defences of irrigation canals, destroying dams, or by diverting streams. It now includes sinking oil tankers,

---

2. See El-Baz (1999) for a comprehensive historical and modern assessment of the environmental impact of warfare, and Kibreab (1997) for an assessment of environmental deterioration in relation to the refugee problem.
3. In 1988 a conflict and poor harvests caused famine in southern Sudan. Aid was denied to tens of thousands of people, and the death rate exceeded 7 per cent per week (Duffield 1990).

thereby polluting the seas and coasts, and bombing oil wells, which pollutes the atmosphere and the land surface. But the increased dependence of the modern world on technology has opened the way to all-consuming **environmental war**, in which the fate of natural landscapes goes hand in hand with the destiny of the protagonists who fight in them (El-Baz 1999). It is not clear whether the staggering US$673 billion spent on the 1990 Persian Gulf War includes the cost of despoliation of the desert surface and seashores. However, there is little doubt about the immense scale of pollution, desertification and damage to dryland, littoral and nearshore ecosystems that the conflict engendered. The future impacts of droughts, flash floods and coastal storms can be expected to be proportionately higher than those before 1990.

Like natural disaster, armed conflict can easily induce a form of dependence on outside aid that is highly detrimental to the long-term survival of victims, or at least to their prospects for wellbeing (Pugh 1998). Roads, tracks and fields that have been wantonly sown with anti-personnel mines become no-go areas or are sources of continual injury. Land, the instruments of agricultural production and food stocks may be expropriated by the military, and labour may be conscripted, even when children are involved. Where insurgency and counter-insurgency alternate, attacks on farms may be common, and in the worst cases farmers are reduced to cultivating the ground under military guard or even at night. Aid dependence comes to the fore when the last thread of resilience has been pulled. Livestock and valuables are exchanged for food until no resources are left to carry food production ahead. If the rural areas are unsafe, their inhabitants may migrate to the cities and swell the ranks of the urban indigent. This is likely to make them politically docile, and to remove a human and economic resource from the territory of the enemy, but it may put great strain on humanitarian aid and create a refugee problem that is difficult to manage (Aga Khan et al. 1986). Part of the management problem may involve negotiating and creating refugee camps that are themselves secure against attack, and also against natural hazards such as floods. Often, only the most marginal land is available for immediate settlement.

The most basic resources in times of flood, drought and military conflict are food and water. Recent years have seen increasing numbers of attacks on convoys of vehicles supplying destitute civilian populations with such commodities. It is no longer safe for the wagons to travel under the banner of some neutral agency, such as the UN Department of Humanitarian Affairs.[4] Soldiers at road blocks may demand bribes by way of toll. Drivers may be hauled out of cabs and shot, and vehicles may be wrecked, hijacked, bombed or strafed.

---

4. A very penetrating discussion of the relationship between military action and humanitarian relief was given by IFRCRCS (1997: chs 1 and 2). See also Stockton (1996) and Weiss (1997).

Sometimes the objective is merely one of obtaining money or supplies, or of turning the convoy back, but in other cases the vehicles have been deliberately destroyed in order to deprive the people in enemy territory of sustenance (Hendrickson 1998, Macrae 1998).

The underlying reasons for this immorality are inevitably complex. Economic mismanagement, corruption and exploitation in high places are tied by complex threads to postcolonial and neocolonial legacies, to the rigours of debt, and often to blatant power brokerage at the expense of the indigenous population. Export-driven economies are needed to sustain high expenditure on arms, but the people who produce the wealth are then deprived of it, which may lead to insurrection and revolt. Separatism and measures to combat it have often tended to nationalize purely local conflicts. In the worst instances, the effective collapse of government has coincided with acute shortages of aid, widespread military instability and severe natural hazards such as drought and desertification.[5]

A resolute stand is needed against these ills. The United Nations has issued various rulings, beginning with Article 3 of the Declaration of Human Rights. They include the UN Convention on Economic, Social and Cultural Rights of 1977, and various protocols and resolutions on the protection of civil populations and on humanitarian assistance in times of armed conflict. Donor countries do attempt to reward nations that reduce their military expenditure, increase civil investment and introduce democracy and transparency into government, but neutrality is very hard to maintain, and the imperative to supply aid is often greatest in the case of nations that are doing none of these things.[6] Moreover, donor countries have tried to use aid selectively as a means of undermining unfavoured governments with or without formal embargoes and blockades. The results have been at best mixed and at worst catastrophic for the hapless citizens of the countries in question (Dammers 1992).

Although the above discussion clearly related most closely to developing countries, and perhaps to some of the states of the Caucasus and Balkans, the world's richer nations have been heavily involved. The rigours of structural adjustment programmes that include hardship and revolt among poor majorities, geopolitical interventionism and superpower rivalry, and the unresolved problems of colonialism all stem from North–South interrelationships with long historical roots. Interventionism continues in both more or less benign forms in the modern era of international aid. In this respect it should be borne in mind that some of the countries that supply humanitarian aid (disaster relief included) also supply the military aid that contributes to the disaster (Pugh 1998).

---

5. Hence the "complex emergencies" described by Duffield (1996) and Kirkby et al. (1997).
6. For example, in Sudan (Macrae et al. 1997).

Military technology

As a coda to this discussion, one other aspect of military endeavour deserves to be examined. As emphasized earlier in this book, technology plays a fundamental role in disasters. If we concentrate on its more benign influence, the extent to which the driving forces have been military rather than civil is instructive. Many of the pieces of technology that are most useful in countering disaster have a military origin. For example, radar, which has been so spectacularly developed for storm monitoring and forecasting, was a military invention. The picture is less clear regarding satellites. Military espionage, surveillance and telecommunications have undoubtedly led the field in satellite research and development, but the technology utilized in Earth resources modules such as the Landsat family is somewhat different and hence has been developed separately (Mack 1990). Nevertheless, much of the justification for the rockets that launch them is military. Moreover, in the early stages of satellite development, there were serious concerns about the effects upon national security of the prying eyes of civilian satellites. The problem was initially solved by making data accessible to all potential customers, but by restricting the character and resolution of images. In essence, technology had run ahead of the military ability to keep it under control.

As technology grows, we see a process of ebb and flow in the strategic sensitivity of environmental data. In the age in which "disinformation" has replaced "propaganda", strategic interests are alternately served and hindered by releasing data on environmental damage and extremes. In this respect there seems to be little difference between countries according to ideology and degree of implementation of democracy. The plea of "national security" is often used as an excuse for furtiveness and dissembling. On the other hand, exposés of other countries' environmental misfortunes and misdemeanours can be used conveniently as justifications for giving or withholding aid.

The logical extension of a consideration of military conflict is to look at the problem of violence. This will now be examined in relation to, and in the context of, natural disasters.

Violence and disasters

Little scholarly attention has been given to the relationship between gratuitous and institutionalized violence, on the one hand, and the inadvertent violence of disaster, on the other. Together with conflict, the former is an expression of human aggressiveness and, at the moment in which it occurs, the latter is unavoidable and inevitable. Let us look at three sources of violence – entertainment, conflict (including violent crime and so-called "recreational violence", i.e. aggressive behaviour) and disasters – in order to see whether there is any connection between them.

Violence as entertainment tends to be either stylized (i.e. ritualistic or symbolic) or realistic, although there is something of a continuum between these two positions. Its extraordinary popularity in the Western world is accounted for by a combination of public appetite, lowering thresholds of tolerance, habituation, and force feeding on the part of the mass-entertainment industry. In some cases (including much popular entertainment), violence on screen is deemed acceptable and even appropriate by people who consider themselves civilized. At the same time, there is an uncertain but very apparent relationship between violence on the cinema and television screens and violent behaviour in the streets and homes of the Western world. Under the corrupting influence of entertainment violence, attitudes seem to have changed towards more passive acceptance, voyeurism, experimentation and involvement, at least among the young and active who have been brought up on a diet of popular video. For those who succumb to its fatal charm, violence corrupts and spreads amoralism. In entertainment violence there is no real concept of fear, no sense of the terribly disorientating effect of uncertainty and unpredictability. Hence, there can be a failure to understand the meaning and significance of violent acts, which partly accounts for the rise in attacks and assaults, if not also for the modern increase in violent crime.

Violence in disaster is a rather different phenomenon, being less orchestrated, unpremeditated and often inevitable. It is also usually a transient phenomenon that is over as soon as the ground stops shaking or the wind ceases to blow. Much of it is unrecorded, as it is seldom captured systematically on film. But it is as traumatizing to those who experience or witness it as gratuitous human violence is. Thus, emergency workers who deal with major death and injury situations often suffer from critical incident stress, a form of post-traumatic stress disorder (Mitchell 1983, Marmar et al. 1996).

Like violence in disaster, the real violence associated with conflicts is not often recorded systematically on film, although it usually has the same impromptu quality. Political and military violence often strike arbitrarily and unexpectedly. Both act as a reminder to the television news viewer that the world is not such a benign place. And both work as an admonishment to improve security, reduce personal risk, and care for innocent victims and survivors. However, both can become a form of voyeuristic entertainment in their own right, fascinating to watch because the viewer is not personally involved and can be thankful for that fact. Whether the impact on television viewers of violence in conflict or disasters inspires prurience or revulsion, it is likely to be similar, irrespective of its source. But, as has been shown elsewhere in this book, conflict and disasters are often linked to each other in real situations (a different way of looking at it is, of course, that conflict is another form of disaster).

The connection between violence in entertainment and in disasters is rather weak. Disaster movies are very popular, but they are almost invariably inaccurate representations, and are often absurd dramatizations. Yet they are the only active visual images of disaster that many people remember. Given the predictable outcomes of such dramatizations, people may think themselves invulnerable and take unwise risks (Turner 1979). But, like most generalizations about the relationship between gratuitous human violence in disasters, there is little empirical evidence. And yet, it is a subject worthy of better understanding, as is the consideration of disasters and aesthetics that will now follow. The two topics are intimately linked in the sense that, according to most moral canons, violence is ugly.

Do disasters make the world uglier?

As beauty is in the eye of the beholder, it is risky to generalize about aesthetic qualities. The short answer to the question is obviously "yes, disasters *do* make the world uglier". Although there are people who would find a certain stern attraction or grim fascination in ruins, by and large destruction does not appeal positively to our aesthetic senses. The splintered beams of houses, the disembowelled carcasses of vehicles, the twisted and torn spans of bridges – their attractions do not amount to beauty, even of a heroic kind. Hence, it cannot be said that disasters create ugliness out of beauty. A telling photograph can be found in one of the US Geological Survey's publications on natural hazards (Brantley 1990). It is a view of Anchorage, Alaska, during the eruption of the Mt Redoubt volcano. A dull blanket of ash covers the entire city and obscures the air, turning day into night. All that shines out amid this murk is a pair of gigantic illuminated golden arches, the "M" of a Macdonald's fast-food restaurant. It is certainly a triumph of corporate aesthetics (and incidentally speaks volumes about the resilience of the hamburger culture), but readers may draw their own conclusions about the values thus upheld.

In as much as the aesthetic values of Western society have undergone a transformation since the Renaissance, there has been a gradual rise in the appreciation of nature and its works. Science, recreation, travel and humanism have all played a role in this. In the Middle Ages – indeed even in the seventeenth century – mountains were feared. Natural landscapes were either accepted in a matter-of-fact way or were loathed and shunned. Only thinkers – Dante on the slopes of Etna, Leonardo da Vinci in the foothills of the Alps – were attracted by the curiousness of the natural spectacle, and their meditations were often afflicted by gloom and pessimism (Alexander 1982a, 1986c).

It was not until the eighteenth century that extreme natural forces began to be admired for their aesthetic appeal: the leaping curve of incandescent magma, the turbulent mud-laden floodwater, the sinuous black funnel of a powerful tornado. In this respect, nature as spectacle probably came before nature as fount of scientific curiosities (Glacken 1967). Yet the painters of the Renaissance, Giovanni Bellini and Pietro Perugino for instance, preferred the orderliness of the managed landscape to the drama of the wild one. Even in Giorgione's *Tempest* the violence of nature is but an intrusion on the "campagnas rich with the blush of fertile Hippocrene" (as Turner put it) that would reach their apogée in the peaceful idealizations of the seventeenth-century French landscape painter, Claude Gellée de Lorrain. Thus, the state of landscape painting remained until the evangelism of the industrial revolution and the cataclysms on canvas of John Martin (1789–1854), painter of the apocalypse, and Joseph Wright, painter of industrial hell-on-Earth. Few were inclined to follow the lead of Leonardo da Vinci, whose science, for so long neglected, merged with his art in studies of deluging water (Alexander 1982a).

According to the canons of modern aesthetics, ugliness resides in the human impact of disasters. In other words, the question of ugliness does not really concern the first effects of disaster, but its ultimate outcome. To begin with, extreme geophysical events tend to "filter out" historical buildings – that is, destroy them selectively.[7] In long periods of quiescence, there is a process of natural selection that tends to favour the survival of the more interesting, worthwhile and opulent buildings (although with many, many exceptions). According to Bernard Fielden, President of the International Council on Monuments and Sites, no historic building need fall victim to disaster, although the cost and technical complexity of saving it may be prohibitive (Fielden 1980).

But the fact remains that historic buildings often present aspects of weakness in the face of catastrophe. First, neglect leads to the propagation of weakness. Cracks in walls, decay of masonry, rotting of beams, all lead to the development of zones of fragility that extreme natural forces will exploit (Hughes 1981, Barbat 1996). Secondly, most buildings are living organisms whose function is adapted to the needs of their users, to the available technologies, and to the level of comfort aspired to by their occupants. At the least, this will involve drilling into the walls of historic buildings to install lighting, plumbing and heating distribution

---

7. For example, see the detailed study of earthquake and landslide damage in a southern Italian town conducted by Rendell (1985).

conduits. In many cases, adaptation has led to alterations in the form and substance of buildings that have made them more complex and have introduced new materials. Thus, in September 1997 the upper basilica of San Francesco at Assisi in central Italy suffered the collapse of its frescoed vaults in an earthquake whose magnitude was only 5.6, because restoration work in the 1950s allowed excessive weight to accumulate on the roof in such a way as to deny flexibility and increase inertial swaying during strong motions of the ground. In earthquakes, buildings of mixed construction and irregular plan tend to perform poorly (Reitherman 1985, Bolt 1993). This presents a good case for strong, simple construction (round towers and rectangular blocks), but it also means that some of the world's most interesting and ornate architecture has succumbed. Finally, historic buildings may be subject to foundation and site constraints that adversely affect their performance under dynamic loading – witness the Leaning Tower of Pisa, which is built on estuarine soil of inadequate bearing capacity, and which resonates alarmingly during earthquakes (Cambefort 1978).

When disaster has struck and buildings have been reduced to rubble, there is often a pervasive desire to create the *tabula rasa* by bulldozing away what remains and rebuilding on a cleared site. It is almost as if the people who are responsible for this strategy blame the stones themselves for their misfortune (Binaghi Olivari et al. 1980). Yet for the most part it is the coward's way out. It rides roughshod over the pattern of property ownership and distribution of urban services, and it sidesteps the issue of whether rubble can be re-used and severely damaged buildings reconstructed. Also, it offers no guarantee that aesthetic standards will be maintained. Indeed, it tends to betoken the sort of haste that almost guarantees that they will not.

The alternative strategy is more complex and may also be more expensive, but it offers a greater chance that the *genius loci* of a place will survive the ravages of disaster. Where historical buildings have been lovingly pieced back together, the numbered stones carefully lifted back into place, there is often a desire to encapsulate the signs of the disaster in the reconstruction as testimony to the event. According to the current fashion in restoration, new stonework may deliberately be given a different character to that which survived the disaster (Vay 1995). The result is an odd-looking patchwork, aesthetically disquieting, a sign of the eternal struggle between harmony and dissonance. We can only speculate on how it will seem to observers far in the future, when time has smoothed over the sharp lines and given a weathering patina to the new stones to match that of the old ones.

Rarely has disaster been used as an opportunity for great aesthetic innovation. Perhaps the reconstruction of Lisbon after the 1755 earthquake is an exception that confirms the rule. It was developmental reconstruction *par excellence*, in the sense of Kates & Pijawka (Kates & Pijawka 1977, França 1983). For such to occur there must be preconditions. In the case of Lisbon, it was one of colonial expansion and the desire to demonstrate power in civic monuments. Ruined by earthquake it was, but Lisbon was still the heart of an empire. The post-earthquake Praça do Comercio, newly reconstructed with shear walls and fire-resistant buildings, opened onto the Tagus estuary as if inviting the world to walk humbly through the triumphal arch at its landward end, and its very name advertises the prevailing mood of that time (Fig. 7.1).

Figure 7.1 Triumphal arch at the entrance to the Praça do Comercio, Lisbon (Portugal), an example of developmental reconstruction (Kates & Pijawka 1977) under the duress of colonial expansionism.

But elsewhere the vicissitudes of disaster tend to lead to reconstruction that is, perhaps, solid but not bold. It shrinks from putting up a fight against the extreme forces of nature. Aesthetically, nothing induces a failure of nerve quite like natural disasters, and it is a failure that is underpinned by shortage of funds, which further constrains the rebuilders.

One curious exception to this rule occurred at Gibellina, a town in the Belice Valley of western Sicily, where in 1968 14 devastating earthquakes left 270 people dead. The ruins of old Gibellina have been transformed into a gigantic environmental sculpture, which is visible from up to 50 km away, a bland white labyrinth that traces out the winding streets of the town that it replaces (see the cover illustration on the paperback edition of this book). The town was reconstructed on a greenfield site nearby and is a bastion of avant-garde art and post-modern architecture. The cryptic metaphors of modernism jostle with constant referrals to Gibellina's Greek, Arab and peasant heritage. Classical theatre is performed in ancient Greek in the local amphitheatre, modern sculptures have been placed on street corners.

In some ways this curious situation, which is the result of energetic promotion of the arts by the town's charismatic mayor, symbolizes the cultural paradox of the modern world. Local people do not understand high art but welcome the initiative as it brings tourist cash to the town. They are not involved in the artistic debate, except with respect to the vandalism that some of the works have suffered (in addition, the parish church, a wildly futuristic cube-and-sphere built in reinforced concrete, collapsed before it could be properly consecrated). In synthesis, the heritage of western Sicily has for centuries been one of marked division between popular culture and elitist high culture, and disaster seems to have exacerbated that, or at least perpetuated it.

On balance, the answer to the question "Do natural disasters make the world uglier?" is a resounding "yes they do". That should be a challenge to the human spirit rather than a signal of defeat.

Anthropomorphism and anthropocentrism

It is a short step from aesthetic considerations to an examination of nature as it is made or conceived in our own image. Like aesthetics, this is relevant to the holistic study of disasters, because a true understanding of disasters cannot be attained without grasping the quintessentially human dimensions of the subject.

A few kilometres north of Florence there is a park, Villa Demidoff, in which stands Giambologna's statue of the Apennines (Fig. 7.2). The sculptor personified the mountains as a muscular, bearded old man, covered with foliage. The

Figure 7.2 Giambologna's statue of the Apennines (1579–80) in the park of Villa Demidoff, Florence.

statue, which was finished in 1580, stands 6 m tall and seems to rise up out of the rocks that surround it, giving an impression of measured and deliberate strength. It is a profoundly artistic example of a ubiquitous trait – the inveterate desire of the human race to depict nature in its own image as a means of understanding and coming to terms with it. So deep-rooted is this tendency that it runs through magic, religion and even science, the three principal ordering systems for human interpretations of nature.[8]

Cultural anthropology is rich in anthropomorphic and zoomorphic interpretations of nature's violence. The angry carp personified earthquakes in ancient China, and in the Pacific the God Pelé would rise up out of the mountain in sheets of volcanic fire (Vitaliano 1973). Whether the deity took human or animal form, it was invested with the whim and caprice, the hidden sense of purpose, of real people – whims to be satisfied according to a long code of accumulated experience and observation of how nature behaves in her extremes. So complex is the accumulated legacy of human experience that magic and religion conflate and colour the popular attitude to science. Hence, the tiny white outstretched arms of the plaster Madonna appear to halt the flow of incandescent lava (although in reality there is usually an element of critical judgement in the placing of the saint). Deep in the Andes of South America the spirits of the Earth are accommodated in the monotheism of the Catholic faith, for in lands where the environment is a hard taskmaster it makes sense to placate more than one God. I recall the peak of a mountain in southern Peru, called Ch'u-Ch'u, which is credited by the local Quechuan peoples as having the power to modify the weather. On top is a Christian cross surrounded by cairns, each one containing a modest offering – a can of sardines, a ballpoint pen. Yet even in societies that like to consider themselves much more sophisticated, nature must be translated into human terms – hence, for example, the way we grace hurricanes and other tropical storms with our own names.

Nevertheless, as society has evolved, there has been a gradual and momentous transfer of principle. Natural hazards – and, as mentioned above, the epithet "natural" is one of mere convenience – have passed from being acts of the spirits and acts of God to being acts of humankind. Inability or failure to impose limits on humanity's use of the environment has made us responsible for the consequences. The result has been the separation of the theological question of God's role in nature's violence from the practical one of humanity's role. The evidence for the latter has accumulated to such an extent that it is no longer possible to shirk the responsibility for disaster by evoking "natural causes" or "divine intervention". The repetitiveness of impacts and forms of damage, the deliberate or inadvertent creation of vulnerability, and the gross predictability

---

8. Meyer-Abich (1997) preferred to call this tendency "physio-centric".

of the consequences of disaster all add up to human, not supernatural, responsibility. This has gradually become encoded in laws, norms and customs, and it represents a fundamental change in humanity's relationship with nature (Beatley 1989).

Responsibility has not always been assumed voluntarily. It poses a question of long-term human adaptability to environmental extremes. This we may classify into four modes, two voluntary and two involuntary. The first voluntary mode consists of unwillingness to adapt and can be termed **hazard aversion**.[9] It involves two possible options. The first is to opt for the maximum protection against hazards; given the very high cost, the benefits must be perceived in order to justify the expense. The second is to migrate out of the hazardous area. The second voluntary form involves a low level of adaptability. Resources are high in relation to losses, and hazard effects are mere imitations (such as the periodic need for evacuation and the occurrence of minor insured losses). The involuntary forms begin with a high level of human adaptability, a state of hazard inversion in which environmental extremes act as absolute deterministic constraints on freedom of action and hence upon the course of human lives. The involuntary form may be termed chronic adaptability, in which the avoidance of hazard is a matter of life or death. Although there are exceptions, the voluntary forms of adaptation are characteristic of the wealthy sectors of industrialized societies and the involuntary forms are mainly associated with poverty and marginalization. Although much knowledge has been accumulated on how poor and wealthy societies adjust and adapt to hazards, much less is known about how the two main forms of adaptation interact. This, of course, is the problem of *equity*, and statistics show that the world has gradually become more inequitable over recent decades (UNDP 1996).

The study of catastrophe is rich in paradoxes. One of these concerns the sanctity of human life and priorities for preserving it. Usually, life saving is the absolute first priority during the immediate aftermath of a sudden-impact disaster, but this cannot be said of the longer term. In part, exposure to a variety of risks means that some loss of life is inevitable. But it is also true that investment and efforts to safeguard human life vary enormously, even within the context of a single type of risk (Foster 1980). The rich will spend millions of dollars on security in return for a handful of lives saved; the poor must all too often fend for themselves in precarious circumstances.

Hazard vulnerability and exposure to risk have much in common with a phenomenon identified as "environmental racism". In this, the "not-in-my-backyard" approach forces hazardous and polluting activities into the neighbourhoods of the poor and politically marginalized, where protests can conveniently

9. Compare this with the wider question of *risk aversion* (Smith 1979).

Figure 7.3 Marginalization of a vulnerable group in a developing country: a floodable squatter settlement on the banks of the River Buriganga in Bangladesh.

be ignored by governments that are dominated by powerful vested interests. From the Americas to Africa, hazardous waste has been conveniently abandoned where it will harm the poor – especially the rural poor – who have little voice in decision making.[10] Similarly, the land that is most susceptible to natural hazards is often reserved exclusively for the poor and marginalized (Blaikie 1985: 125). Steep tropical hillsides with unstable slopes, flash-floodable canyon floors, coastal sandbanks subject to storm surges – all are the domain of the poor and dispossessed (Fig. 7.3). This alone may account for the glaring disparities in death tolls between industrial and agrarian societies. Landslides in a developed country of medium size with a population of, say, 60 million inhabitants are unlikely to take more than about 40 lives in an average year. A single catastrophic slope movement in a precariously urbanized *barrio* on the fringes of a large tropical primate city may well take more lives than that (Jones 1973, Jibson 1989, Kelly 1995), and it is unlikely to be considered an exceptional event.

In synthesis, anthropocentrism is now driven by the force of numbers (population trends, that is) and mass expectations, and by frantic attempts to satisfy the latter. But expectations vary markedly in terms of who expects what from whom. Morality demands that we move to restore to human life some equity in the ability to adapt to hazards.

Now we turn from humanistic issues to science, their counterpart. One philosophical question begs to be tackled and it is that of classification, a subject that has as many arbitrary perceptual connotations as it does objective scientific ones. Although it may be imagined that taxonomy became a cut-and-dried

10. See Clarke & Gerlak (1998) for a critical analysis of this phenomenon at the local scale.

process in the eighteenth-century world of scientific discovery, it has instead undergone a recent upheaval: the welter of information that bombards the citizen today demands some form of rapid classification in order to make it comprehensible. But lack of overt attention to the categorization *process*, and general superficiality, can result in taxonomies that have little validity. Classification is also a basic inductive process in the study and management of disasters. But even here it is not immune to the trends and fashions that prevail in modern society. Some of the same cultural and perceptual influences are at work as we find in anthropomorphism (Barker & Miller 1990). By failing to identify what is genuinely important in causing disasters, it is hard to classify them and their effects successfully. The wheel of science and society turns full circle: society's preoccupations become those of science and vice versa. In short, the classification of disasters and their effects is not quite such a solved problem as one would expect after the better part of a century of the intensive study of natural catastrophes. The next section will examine why this is so.

Classification, taxonomy and ranks

In modern society there is a pervasive desire to classify and to relate things to each other. It owes less to the eighteenth-century researches of Carolus Linnaeus than to the current quest for certainty in a world that changes ever more rapidly. We feel impelled to rank the unrankable, to assign numbers, often quite spuriously, to intangible values, to fit concepts, however awkward they are, into a simple scheme. In this respect, classification, the pigeon holing of phenomena, should be distinguished from taxonomy, which fits them into a more powerful hierarchical schema. The former is a low-level kind of non-hierarchical inductive model, whereas the latter may contain elements of deduction and refer to process–response systems. But in any case, the pervasiveness of the modern desire to rank things is partly a reaction to the technological mechanization of life and partly a forlorn desire to render more homely the constant stream of information that bombards us.

The study of disasters has been dogged by the problem of classification, as one of the fundamental needs of such a young field has been to impose order on the phenomena under examination. It has not proved easy, as natural catastrophes are multifarious events in which uniqueness vies with regularities that fit into at least 30 of the traditional academic disciplines, and this offers a wide choice of perspectives. But it may be that, in this field, taxonomic processes have been rendered more complicated by some of the attitudes to classification that have developed in the modern world. We will now examine these in some detail.

189

The antecedents of the problem lie in the rise of modern advertising in the 1950s, when every product was either "the best" or "better" – the *metacomparison*: faster, sweeter, more, but in relation to nothing. The message has been beamed at us for so long now that we have become inured; it seems normal that something can be better without being compared to anything.[11]

It is an obvious fact that not everything or everyone can be "the best", and hence we rank, or we assign indices and percentages. The worth of a book, or a college, or a service, is determined by its rank: "number three on the bestseller list", "the fourth best in the nation", "an impact factor of 5.3", and so on. We have convinced ourselves that the rankings are true, even if they last for but a single day. It always amazes me that intelligent people will preface a conversation by saying "of course, I don't believe it means anything, but . . ." and then go on to demonstrate a sincere belief in the shakiest of indices or rankings. They adhere to these values like shipwrecked mariners clinging to flotsam and jetsam amid the stormy seas of life's essential uncertainty. And so with disasters, we rank the "worst", the "greatest", the "most serious", but the ranking does not really confer meaning, significance or gravity upon that which it classifies.

This tendency also has origins in psychometrics, empirical social research and physics. The first of these gave us the intelligence quotient and the pervasive myth that by using a single number human intellect could be pinned like a dried butterfly to a card. The popular equivalents are the quiz show, in which so-called intelligence is measured competitively by a person's ability to memorize information and regurgitate it uncritically when cross examined, and the test of "scholastic aptitude". In the Western world we now have more than 50 years of experience with aptitude tests, but no-one can say whether they truly measure aptitude or whether, at least in a sizeable minority of cases, they suppress it.[12]

Social research has given us the opinion poll, much used by sociologists of disaster and by students of the perception and psychology of catastrophes. At its crudest it demands a yes or no answer to questions that so often require a caveat ("yes I probably would, but only if . . ."). The result is a truncated form of reality: "33 per cent of people believe that . . ." – perhaps they do, but under what circumstances, and for how long? The most remarkable aspect of this is the way that opinion polls have been turned into self-fulfilling prophecies by

---

11. Thus, the alarm bells rung by Vance Packard in his 1950s book *The hidden persuaders* (Packard 1981, a classic critique of advertising) have largely been ignored. His death in 1997 was greeted with a resounding silence on the part of the chattering classes and the mass media he criticized so aptly. Perhaps that was their revenge.

12. A positive view of aptitude tests was given by Lindvall & Nitko (1975) and a negative one by Crouse & Trusheim (1988). In the years that separated these works the worm may have turned: psychometric measures of educational attainment are no longer as popular as they were in the 1970s.

using them as a form of positive feedback to influence future opinions. If the current poll does not produce the desired result, then commission another and another until public opinion, swayed by the figures, comes around to the preferred way of seeing things.

In history, the balance of attitudes has tended to swing from embracing uncertainty to rejecting it. In the twentieth century, physics and mathematics reclaimed uncertainty after the moralistic determinism of the 1800s. Moralistic determinism enjoyed a brief swansong in the first three decades of the twentieth century under the aegis of colonialism, although curiously it was not subverted by fascism (Holt-Jensen 1988). But then it perished amid the welter of stochastic processes. Parametric statistics were born in a field at Rothamsted Experimental Station (Harpenden, England), where the likelihood that each head of turnip would survive and grow had to be ascertained. The parameters were the 90%, 95% and 99% confidence limits that were arbitrarily used to assess the significance of the results of the tests. In physics, Werner Karl Heisenberg (1901–76) showed that it is impossible to make a precise, simultaneous measurement of both the position and the momentum (mass × velocity) of a body. Stochastic processes (from the Greek *stochastikos*, to aim or guess) developed out of this, and perhaps so did chaos theory. But paradoxically all this has not enamoured us of chance. Uncertainty has been nailed to a board, pinned down as probability, likelihood, confidence levels; we have no theory of improbability, unlikelihood or lack-of-confidence intervals – we do not look for such things. Even chaos theory seeks order in confusion. Moreover, Heisenberg showed that his uncertainty principle boiled down to an immutable value (Planck's constant). Although such matters seem far removed from the more humble deliberations of workaday life, it is not so, for the backwash of science permeates the attitudes of people who are utterly unaware of its precepts. Thus, in our efforts to comprehend disasters we must cope with some final and essential uncertainty, which is the enemy of classification, for it undermines judgement and makes classes and labels unreliable. Moreover, indeterminacy is the enemy of legitimacy, which is what a young discipline will tend to seek. It reduces apparently solid generalizations to the level of speculation, or at least to partial truths. The problem is particularly evident with regard to the social study of disaster, in which most generalizations seem to be countered by abundant exceptions. Quantification may add to the social taxonomist's difficulties, rather than reduce them, because it tends to create a spurious air of accuracy and reliability.

Quantification, classification and taxonomy are not synonyms, but they are complementary procedures. Any one of these legitimizes, or at least consolidates, the others. Taxonomies may consist of classes, even if the latter can exist independently of the former, and classes may be defined numerically. In one sense they are the three pillars of the self-deception that is inherent in so much of modern thought. They are low-level models that flout the principle of modelling, for, rather than encapsulating phenomena by elegantly simplifying them, they force awkward concepts into a predetermined mould; they *impose* shape and form rather than identify it. In another sense, numbers, ranks and classes are our feeble defence against the great threat of unknowability. We can handle "nothing is what it seems", for that is a challenge, but we cannot cope when "things seem not to be anything".

In this, disasters can be considered to be the harbingers of change and chaos, phenomena with ragged edges, upsets and perturbations, shocks to the system (Alexander 1995a). Although each catastrophe may be unique in its details, there is a fair degree of aggregate predictability. Regularities are easy to cull from disasters, hence the cycle depicted in Figure 1.1. But classification is another matter. Many attempts have been made to quantify disasters, and to invent classifications and taxonomies. I must admit that it is a lure to which I am far from insensitive. Yet most disaster taxonomies are either facile or inoperable. What should they be based upon? Numbers of deaths and injuries? The dollar value of damage? The sum total of human misery? No combination of factors is without snags.[13]

To begin with, there are few real thresholds. Even the physical side of disasters has to face this problem. For instance, the lowest windspeed of hurricanes, 65 knots (120.4 km/hr), is an entirely arbitrary figure (Alexander 1993: 155). On the human side there is even less to hang on to. If I am killed in the next earthquake, then as far as I am concerned the death rate will be 100 per cent. In the second place, there is little justification for adopting internal thresholds for most scales of damage or casualty: in process and magnitude terms it is arbitrary to distinguish between an "incident", a minor disaster and a major catastrophe. Finally, it is well known that physical and human measures of disaster tend to coincide very poorly. Thus, a magnitude 7.5 earthquake deep within the Earth's crust in the Tonga Trench (in the South Pacific Ocean) may hardly be felt at the surface, whereas a shallow-focus tremor of a similar magnitude may cause many deaths if it occurs beneath a major city.

In terms of taxonomy, disasters are characterized more by continua than by thresholds. Between 20 seconds of catastrophic earthquake shaking and centuries of accelerated erosion there is a wide range of timescales of forewarning, impact, emergency and response (Fig. 7.4). Disease rates provide a yardstick for natural disaster risk analysis,[14] but there is considerable uncertainty about whether they should be included as natural disasters themselves, and likewise

---

13. See pp. 36–39.
14. For more details on this, see the following works: Petak & Atkisson (1982), Burton & Pushchak (1984), Smith (1996: 54–77), Tobin & Montz (1997: 292–302).

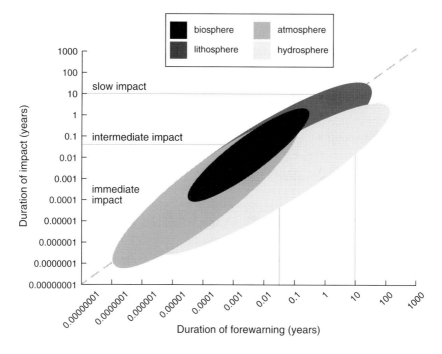

Figure 7.4 A classification of disasters based on duration of forewarning and impact (after Alexander 1994a: fig. 4).

for epizootics and infestations. The fashionable view that the causes of natural disaster lie more in human organization than they do in geophysical phenomena (Hewitt 1983, Quarantelli 1995) calls into question the distinction between natural and technological disaster.

Natural disaster studies have been seriously inhibited by the difficulty of imposing a rigorous classification and taxonomy, for these are the hallmarks of a fully fledged science. However, we should not try to reinvent reality as something that is simple and linear. Instead, we should have the courage to seek truth in complex situations and, where necessary, to sacrifice generality to uniqueness. Much of the fascination of disasters lies in their degree of unclassifiable uniqueness, for the plight of victims and survivors cannot be minimized by generalizing it. Thus, our response to the tendency to rank and assign numerical values should be to buck the trend: we should search for means of analyzing numbers of events that preserves, rather than destroys, their essential uniqueness and their freedom from comparison.

The next chapter will consider aspects of natural disasters in developing countries. In many respects all societies are in transition in a time dominated by changes as dramatic as those that are currently occurring, but the process is particularly striking in parts of the Third World. At the same time, the majority of academic literature on natural disasters has been produced by, about and on behalf of the industrialized countries. Thus, we observe not merely a widening development gap but also a widening perceptual abyss. It is as if the world's poorer countries had no sociology and only a rather rudimentary cultural anthropology. In reality it is not so, but few sociological researchers have had the curiosity and the wherewithal to apply sociological methods to such societies, many of which are highly developed but upon lines that do not intersect easily with those of the industrialized world and thus do not fit easily into the generalizations of the developed world's sociologists. In any case, the changes currently being wrought in the developing countries are certainly no less profound than those in the richer nations. Moreover, the Third World bears some 90–95 per cent of the deaths in natural disaster and perhaps 80 per cent of the overall impacts, depending on how these are measured (Blaikie et al. 1994).

CHAPTER 8

Worlds apart

One of the most striking aspects of the modern world is the persistent gulf between the rich and poor nations. It is a deeply rooted phenomenon that has more or less prohibited any convergence between attitudes and expectations. Old-fashioned colonialism may have ended, but that has not allowed the world's poorer countries to grow more like the richer ones. Other forms of exploitation, often cruder and even more destructive, have taken the place of colonial overlordship, and neocolonialism has come to the fore in the depredations of capital and commodities markets. Most depressingly, warfare has continued unabated in countries that desperately need stability in order to create economic growth and wellbeing. Often it has been conducted as a proxy for confrontation between global or regional powers; at other times it has resulted from more direct exploitation and repression. Thus, it is hardly surprising that attitudes to hazards, impacts of disasters and strategies to mitigate risk all differ substantially between the rich and poor countries. However, this is not to imply that all Third World countries are alike in their response to natural catastrophe (Wijkman & Timberlake 1984). Especially in this case, "Third World" is a rather misleading term, as it masks a wide variety of problems and solutions. This chapter will delve into these in order to investigate what disasters really mean to the developing world and what light they throw on the tortuous relationship between the rich and poor countries.

The first few sections of this chapter will offer a survey of the current situation regarding disasters in selected parts of Asia, the Middle East and Africa. Following on from observations about disaster relief made in these sections, a brief study of World Food Programme (WFP) operations will be presented in order to illustrate some of the dilemmas of international aid in the modern world, when military action commonly affects disaster aid and the relief agencies find themselves at a moral and operational crossroads (Hendrickson 1998). In the interests of holism, the two sections that follow this will change the focus from the international to the local level, in order to assess how disasters appear to the people who matter most: the potential and actual victims. Finally, the conclusion of this chapter will briefly assess the prospects for disaster relief in developing countries. It will also tackle the vexed question of whether interest in the poorer countries' disaster problems on the part of the richer nations can be sustained, given the occurrence of "donor fatigue" (Macrae 1998).

Natural disaster in Asia

In a world context, Asia stands out in terms of the impact of natural disasters, in both historical and contemporary contexts (Freeberne 1962, Uitto 1998). Poverty, high population densities, and the distribution and variety of natural hazards combine to produce situations of persistent and recurrent tragedy. To understand this situation better it is worth examining the particular vulnerability and recent record of catastrophe in several Asian countries. The choice is an open one, as many criteria justify the selection of different nations, but let us focus on China, the Philippines and Bangladesh as examples of Asian vulnerability. In an account such as this, it is impossible to survey the situation in each country thoroughly, but it is possible to examine why vulnerability should be so high there and to develop a comparative perspective on each case.

China

In many respects, China is the country that is most affected by natural disaster. Some 1156 million people live in its 9 536 500 km^2 and have a mean per capita GNP of US$370. The vast land of China is climatically and geologically of extreme diversity and it varies from uninhabited desert to some of the world's largest and most densely settled conurbations. Every kind of natural hazard is present, and records indicate that serious natural disasters have occurred in at least one province, and sometimes as many as a dozen, virtually every year for the past two millennia (Wang & Zhao 1981, Jones 1987).

One of the most widespread and serious problems that China faces is land erosion and degradation, which has a strong impact on the nature and incidence of floods, landslides, subsidence and desertification. Some 150 million ha, or 15.6 per cent of national territory, are eroded, which represents an increase of 30 per cent on 1979 values (Han Chunru 1989, Wen Dazhong 1993). Soil erosion is estimated to have increased by 100 per cent in the past three millennia and 50 per cent in the twentieth century alone. Two thirds of the 720 000 km^2 Yangtse River basin is suffering from erosion. It is particularly serious in the Loess Plateau, an area of 530 000 km^2, which is covered by deposits of aeolian silt

30–200 m thick. Sixty million people live in this area and one third of soil loss comes from the land that they cultivate. Three quarters of the plateau shows significant signs of erosion, which yields between 20 and 200 tonnes/ha/yr of sediment, a total of 2.4–2.5 billion tonnes per year. Much of the solid load goes into the Yellow (Huanghe) River, as the Loess Plateau falls within its 680 000 km^2 watershed. The sediment load at the mouth of this 5464 km-long river, which flows for 400 km across the loess deposits, is estimated to be 1.2–1.6 billion tonnes/yr (equivalent to 1 cm per year of soil loss across the eroded area of the plateau) and deposition along the lower reaches has raised river-bed levels by up to 12 m above the surrounding land. When floods occur, they tend to inundate such vast tracts of land that tens of millions of people are affected, and damage, especially to crops and villages, amounts to billions of dollars (Robinson 1981, Jing Ke 1988).

Seismically active fault complexes and neoseismic orogens abound in China, which has some spectacular tectonically dominated landscapes (Doornkamp & Han 1985, Petrov et al. 1994). Records indicate that the largest-known earthquake death toll (830 000) occurred in 1556 at Shensi. Indeed, China accounts for about half of known earthquake deaths, owing to the combination of high recurrent seismicity, high population densities and weak building stocks (Coburn & Spence 1992). In the 1976 Tangshan earthquake, in which 240 000 people died, only four multistorey buildings survived relatively intact out of 352, and a city of 1 million inhabitants was reduced to rubble (Cheng Yong et al. 1988). Earth tremors are, moreover, frequently linked to mass movement; for instance, the 1920 Gansu earthquake caused flow failures 1.5 km long that killed 100 000 people (Song et al. 1989). In all, at least 156 lethal earthquakes have killed 610 000 people in China in the twentieth century (Chen Zhiming 1996).

Many other hazards afflict China. Like floods, drought is an annual occurrence and in historical times it has resulted in about 5 per cent of the population dying of starvation. Desertification is also serious: of the 11 per cent of the country that is under crop land, one fifth suffers from salinization, and in the north and west of the country migrating sand dunes are a problem over vast tracts of land. In the Chengdu area of Sichuan Province massive rock avalanches and debris flows are a persistent problem, because of unstable mountainous terrain and a wet climate (Bruhn & Li 1989). In the South China Sea, typhoon tracks frequently bring major tropical storms to landfall along the populous coast, with inevitably high numbers of casualties and high losses (Beijing Normal University 1992).

The vicissitudes of the Chinese people have been numerous during the Revolution, Cultural Revolution and subsequent economic liberalization. China has remained throughout this time a country of tensions between provincialism and bureaucratic centralism, a polarization that has characterized it throughout history. But attitudes to natural disaster have altered significantly in recent decades. The violence of nature is no longer seen as a disgrace to be concealed lest it reflect in an unfavourable light upon economic progress (Fan Yuchen 1991). At the same time there has been an increasing desire on the part of the Chinese to learn from the rest of the world, to share data and experiences and to participate in global initiatives to reduce disaster impacts. There has thus been a shift of emphasis away from the previous reliance on autonomy towards collaboration.

Nevertheless, despite the strengthening of scientific, technological and organizational defences against catastrophe, the principal resource remains human. The high degree of socialization in Chinese society, and a traditional emphasis on obedience and authority, enable tremendous collective efforts to be made in times of emergency, where in similar circumstances Western societies would rely on much more selective intervention in a context of greater individualism. But the human resource cuts both ways, for it is widely held that population control, which is proving difficult, is the key to land protection, especially in the loess belt.

There are also signs of the emergence of differential vulnerability. Most of China remains a poor agricultural country, where rural systems bear the greatest impact of disasters (Go Huancheng et al. 1989, Han Chunru 1989). This is as it has been for millennia and progress in civil protection has been muted. But the so-called new economic zones are beginning to demonstrate another kind of vulnerability based on the coincidence of high population densities, concentrated capital assets, uncontrolled industrial hazards and recurrent natural hazards (Lu Jingshen et al. 1992). It is much the same as the flood, typhoon and landslide hazard in populous Hong Kong (Lumb 1975, Brand 1994), which is of course in the process of integrating with the Chinese territories farther north. In synthesis, this new differential poses a serious dilemma in the apportionment of national resources for disaster mitigation, reduction and management. The persistent state of ferment in China gives little indication of how this quandary will eventually be resolved.

The Philippines

In the Philippines archipelago, about 62 million people live in a land area of 300 000 km^2 at an average density of 207 people per km^2. GNP is relatively high at US$740, but 49 per cent of the population live below the poverty level and income differentials can be very high. The most damaging natural hazard that the country faces is the typhoon (Fig. 8.1). As 38 per cent of all major tropical storms originate in the northwest Pacific Ocean, an average of 20 typhoons affect the Philippines each year and 10 actually cross the country (Pineda 1993). In 1993, however, 32 typhoons arrived. Hazards result from high winds, storm

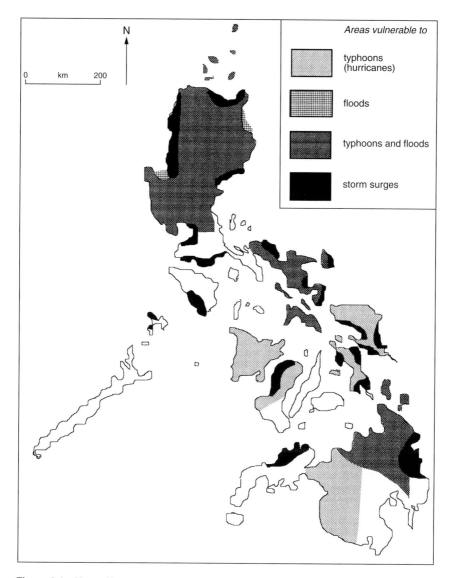

Figure 8.1 Natural hazards in the Philippines: areas of impact of typhoons and floods (after Pineda 1993).

surge, floods and landsliding. Death tolls tend to be relatively limited (although they are not infrequently in the hundreds), but the constant repetition of cyclone impacts has a cumulative effect in terms of human suffering, damage, displacement, crop losses and land degradation. In connection with the exceptional number of cyclones, 1993 was also the wettest year on record, with 17 parts of the country affected by floods. Previous episodes of flooding have also been catastrophic. For example, in 1972, floods in central Luzon, the most populous part of the Philippines, killed 600 people, left 370 000 homeless and damaged 250 000 ha of crops. The combined average death toll in cyclonic storms, earthquakes and floods is 500 per year, but the Mindanao floods of 1981 alone killed 200 people. On the other hand, in 1987, drought ruined 41 900 ha of rice and corn, and damaged a further 141 700 ha. Two years later, in 1989, 31 500 ha of crops were affected by drought.

An extreme climate is complemented by extreme geotectonic problems in the Philippines. An average of 16 perceptible earthquakes occur each year, with occasional disasters; for example, the Mindanao tremors of August 1976 killed 3792 people and the Luzon earthquake of 16 July 1990 (magnitude 7.7) killed 1666 and injured 3561 (*Earthquake Spectra* 1991). In the twentieth century 10 200 people have been killed in 25 major earthquakes. Offshore and Pacific-floor earthquakes also cause disasters in the Philippines, especially in the coastal area of Mindanao facing the Celebes Sea, and by generating tsunamis (Hatori 1996). From 1603 to 1976, 27 damaging tsunamis were recorded. On 16 August 1976 a 5 m tsunami killed 3000 people, injured 8000 and left 12 000 homeless. In addition, 18 volcanoes are believed to be active in the Philippines. Eruptions occurred in June 1991 on Mount Pinatubo (Pinatubo Volcano Observatory Team 1991, Wolf 1992) and in July 1996 on Mount Mayon, but Taal, Hibok-Hibok, Bulusan and Canlaon are also particularly active (Fig. 8.2). Besides ashfall, volcanic mudflows (lahars) are a particular and recurrent problem in the populous areas surrounding volcanoes such as Pinatubo (Pierson 1992). Lahars can travel far and rapidly, cause major destruction, disrupt drainage and lead to flooding, and they can occur both during and in the absence of eruption.

Lahars are one category of mass movement, or landslide, a class of phenomena that is particularly important in the Philippines and which relates to a variety of other natural hazards. Landslides are ubiquitous on Luzon and are especially common in the Baguio area to the north of the island, where elevations exceed 1460 m (Arboleda & Punongbayam 1991). As one example of a landslide disaster, in 1985 the gold mining areas of southern Mindanao suffered unusually heavy rains, and 400 miners and their families were buried by slope failures. The July 1990 earthquake was followed one month later by a major typhoon that augmented pore-water pressures in slopes that had been destabilized by the tremors and aftershocks. Landsliding was particularly widespread and serious in Baguio and Nueva Viscaya. But mass movement is also part of a more general problem of soil erosion. Some 13 Philippine provinces show signs of accelerated erosion in more than half of their area and about 22 per cent of existing and potential farmland (which comprises 13 million ha) is susceptible to erosion (Broad & Cavanagh 1993).

In synthesis, the Philippines are highly vulnerable to a wide variety of natural

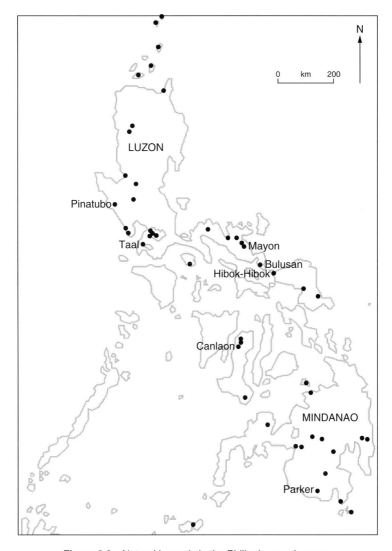

Figure 8.2 Natural hazards in the Philippines: volcanoes.

catastrophes. Two thirds of the population (40 million people) live in the seven regions that are most susceptible to typhoons. Crops are vulnerable to floods, high winds, the accelerated erosion of topsoil, landslides and effects of volcanic eruptions such as ashfall. The fishing industry, which is both intensive and highly competitive, is vulnerable to typhoons, storm surges and tsunamis. Inland roads are particularly susceptible to flood and landslide damage, and unreinforced evacuation centres have suffered roof collapse with casualties as a result of high winds or the accumulated weight of volcanic ash (Tayag & Punongbayan 1994).

In the Philippines the central question of mitigation and protection is whether they can be achieved with limited resources. The National Institute of Volcanology and Seismology (PHIVOLCS) is highly active in monitoring volcanoes and seismogenic areas. As a result of this and well organized civil protection, evacuation before the June 1991 Pinatubo eruption was comprehensive and efficient, and the system of alert levels used was the right one (Pineda 1993). The situation invites comparison with that other vulnerable archipelago in the western Pacific, Japan. Both suffer from the combined effects of tectonic, volcanic and erosional activity, and extremes of weather. Both are populous (the Philippines has an average population density of $207/km^2$ and Japan $332/km^2$) and both have a strong imperative for civil protection. But there is a major difference in the distribution of monetary and technological resources. Japan can afford to rely heavily on the deployment of high technology, but the Philippines must depend very largely on resources of organization. This may help to reduce casualties somewhat, but it does little to reduce damage. In the future, increasing population pressure on land and environment, and increasing competition for resources, will stretch the inventiveness of civil protection planners. The success of mitigation efforts will probably depend in large measure on what effect economic policies have on population growth, poverty and general vulnerability.

Bangladesh

At $144\,000\,km^2$ Bangladesh is slightly larger than England and approximately the same size as Wisconsin, but it has a population of 120 million, with a density of more than $850/km^2$, one of the highest national totals in the world. Moreover, its 8 per cent urban growth rate is almost twice that of the Third World in general and so the country can be considered a veritable "demographic bomb". Four fifths of Bangladesh consists of the deltaic floodplain of a remarkable complex of 254 rivers. Some 56 originate outside the national borders, including the Ganges (Padma), Brahmaputra (Jamuna) and Meghna, which have a combined basin area of 2 million km^2, only $60\,000\,km^2$ of which is in the delta (Brammer 1990). The Ganges–Brahmaputra system has an annual sediment discharge of 2.5 billion tonnes and is very mobile in its lower reaches: lateral shifts can be as great as 800 m per year (Coleman 1969, Latif 1969, Bristow 1987).

Between 1960 and 1981, 63 disasters occurred in Bangladesh with the loss of 655 000 lives (Alexander 1993: 532). Of these, 37 were tropical cyclones, which killed a total of 386 200 people. Death tolls have remained significant and occasionally high since then at about 40 000 a year and, even when they have been limited, the number of people affected by flooding has often been very large, as in the case of the 5.67 million impacted by the July 1996 floods. Moreover, although earthquakes have killed very few Bangladeshis in the twentieth century, the country lies near some powerfully seismic areas of the Himalayan foretrough and foothills (Fig. 8.3). The experience of the northeast Indian earthquakes of 1757 and 1934 is that, through the phenomenon of liquefaction,[1] they have the ability to modify drainage very considerably in Bangladesh, and indeed to re-route major rivers (Hoque & Khandoker 1990).

Floods can be divided into seasonal and contingent varieties. In an average year, seasonal rainfall and river floods may inundate 20 per cent of national territory, and in exceptional years contingent flooding can increase the area to 30–40 per cent. Although the reliability of the data has been questioned (Rasid & Pramanik 1993), it has been estimated that the floods of July–September 1988, which resulted from an unusual coincidence between the peak flows on the Padma and Jamuna rivers, put up to 58 per cent of the country under water. Thirty million people were left homeless and 1657 were killed. Normally, however, rainfall floods are as important in inland Bangladesh as are river exundations. Along the northeastern and eastern flanks of the hill tracts, flash floods are often highly destructive (Brammer 1990).

But the main source of mortality comes from tropical cyclones. The 1970 cyclone caused a 7 m storm surge that left 85 per cent of families without shelter on the coastal plains of Bangladesh (especially in the mobile but settled offshore sandbanks known as *chars*) and killed an estimated 224 000 people. Two thirds of coastal fishing enterprises were destroyed (Islam 1974). Subsequently, the May 1991 cyclone killed 145 000 people and caused about US$1385 million in damages in the Chittagong area (Khalil 1993, Mushtaque et al. 1993).

Not only is population highly concentrated in Bangladesh, but there is substantial overlap between the areas affected by each natural hazard (Fig. 8.4). Physically, the impact of flooding is being worsened by subsidence along the Lower Ganges corridor and in the Sylhet trough in the northeast of the country. It may also be significantly worsened if sea level rises as a result of the greenhouse effect, which could also drive salinity fronts (where seawater interfaces with river water) inland into prime farming country. Bangladesh is an overwhelmingly agricultural country with 68 000 villages and nearly three quarters of employment in farming and fishing.

The problem of natural hazards in Bangladesh is a multifaceted one in which a key role is played by poverty and associated vulnerability. More than 100 international aid agencies operate in Dhaka alone, but they cannot solve the problem, only partially alleviate it. Given the ever-present problem of hunger

---

1. Liquefaction is the spontaneous change of a saturated granular or clayey sediment from a solid, which may be able to support buildings, to liquid behaviour, including the production of sand boils (small fountain-like structures) and spontaneous surface collapse. It often results from earthquakes (Committee on Earthquake Engineering 1985).

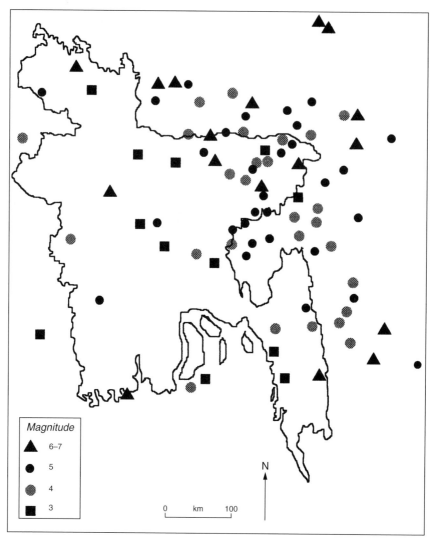

Figure 8.3 Earthquake epicentres over the period 1663–1989 in Bangladesh.

and possible starvation, a crucial factor is self-sufficiency and equitable distribution of food. With regard to the production of basic foodstuffs, mainly varieties of rice, Bangladesh's fortunes have oscillated. Self-sufficiency has briefly been achieved, but it depends on a combination of political stability, social peace in the countryside, and freedom from major disaster impacts (Brammer 1990). That mixture is hard to sustain, especially as fertilizer use is very low in Bangladesh and experience with high-yield varieties of rice (products of the "green revolution") has been rather mixed.

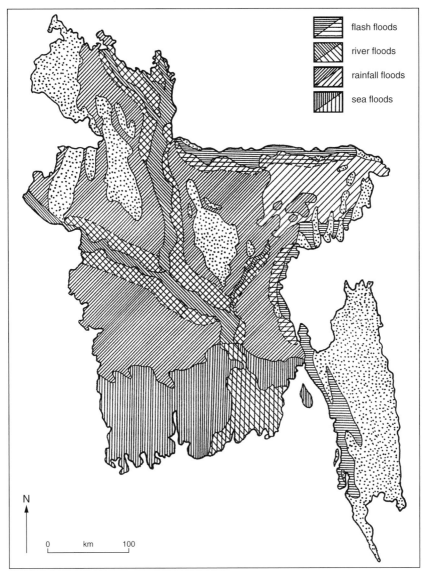

flash floods

river floods

rainfall floods

sea floods

N

0   km   100

Figure 8.4 Areas of flooding in Bangladesh.

Considerable progress has been made in monitoring and warning. It is now feasible to predict inland floods two to three weeks in advance, using a combination of regional hydrological and satellite-based meteorological monitoring. At the same time there have been significant improvements in the organization and practice of civil protection. Warning and evacuation were absolutely inadequate in the 1970 cyclone disaster, but were well organized in the 1991 event. By the end of 1994 Bangladesh had built 1275 cyclone shelters for coastal evacuees and many of these buildings have space for up to 1750 people. They also function as schools, clinics and community centres during normal times. However, many more are needed.

Besides the natural and global anthropogenic causes of environmental decay in Bangladesh, there are social and demographic pressures that contribute to the problem. Deforestation is more or less complete on the floodplain and is accelerating in both the Sundarbans mangroves and the Chittagong Hill Tracts. Landownership and land reform remain fiercely disputed issues and exploitation of the poor is widespread. Moreover, the prospects look relatively poor, as many such issues are not being resolved. The dispossessed flock to the cities, especially Dhaka, where a quarter of the inhabitants are reckoned to be the victims of rural disasters, both natural ones and those induced by land possession issues. However, there is some evidence that democratization is giving rural people a greater voice in their own affairs and especially in questions of disaster mitigation, although progress is very slow (Haque 1988, Zaman 1989, Nurul Alam 1990, Paul 1999).

In Bangladesh, domestic political issues cannot be resolved without reference to the broader regional environmental situation in the Indian subcontinent (Kattelmann 1990).[2] Since 1974 India has diverted the waters of the Ganges by the Farakka Barrage, into the Hooghly River, in order to maintain a navigable waterway to Calcutta. Despite the existence since the 1970s of a bilateral treaty for water management, Bangladesh has remained furious but impotent in the face of blatant hydrological manipulation conducted only few tens of kilometres from its northwestern border. The Farakka Barrage has little impact on the flood propensity of the River Ganges but a great effect on the converse problem, drought. It therefore puts the lives and livelihoods of millions of Bangladeshis at risk. One hopes that the implementation of a new shared-waters treaty will reduce the problem, but many Bangladeshis are sceptical.

Both India and Bangladesh have proposed engineering solutions to the problem by means of constructing link canals. These plans need to be evaluated in the context of a powerful debate about engineering approaches to a flood problem that has gone on for decades and has been particularly intense for the past ten years or so. Given the success of canalization on northern Europe's major rivers, Dutch and French engineers originally proposed the containment of the Padma and Jamuna Rivers in Bangladesh, using a system of levees that would allow water and sediment to be poldered in the wet season for farming purposes (Rasid & Mallik 1993). The plan was based on the partial success of both poldering and embankment schemes in various parts of central Bangladesh. At

---

2. A new accord on the sharing of Ganges water was signed in 1997, but the extent to which it will resolve the dispute between India and Bangladesh is not certain.

the height of the original Flood Action Plan, in the mid- to late 1980s, high-level political promotion by France, strongly favouring the use of French expertise and contracting companies, led to a World Bank Scheme to fund the project by up to US$10 billion. The result would have been a catastrophic waste of money (Boyce 1990, Westcoat et al. 1992) and the same can be said of the link canals proposed to alleviate the effects of the Farakka Barrage. For Bangladesh is geologically, hydrologically and environmentally very different from the plains of northern Europe. Its rivers are much more powerful and mobile, their sediment loads much higher (Alexander 1994b). The Ganges–Brahmaputra delta is de-watering and sinking as it compacts under its own weight. It can maintain itself only by distributing sediment through the lateral mitigation of the rivers. This is achieved by the vast forces inherent in helical flow (the rotational effect that causes meander bends), the Coriolis effect and the dynamics of flood scour. Human use of the land must inevitably adapt to this and any attempt to thwart it is likely to fail. In the end, levees and embankments may be breached, not only by unusually strong flood currents, but also by earthquake-induced liquefaction (Hoque & Khondoker 1990).

Seen in perspective, and in the light of demographic and economic trends, the problems of Bangladesh look particularly intractable. This may be so, and the country undoubtedly has little option but to export as much of its labour as it can, a process that has proved especially difficult in the aftermath of the Gulf War of 1990 and the repatriation of 90 000 Bangladeshi foreign-wage earners. But there is still much room for human ingenuity in the way that flood, drought and cyclone problems are tackled in Bangladesh. Large-scale infrastructure projects have shown themselves to be redundant in many cases, but there is immense scope for modest forms of technology transfer.[3] As Bangladesh is a perpetual recipient of international aid, on which it remains chronically dependent, there is very good scope for the sort of partnerships in development that accomplish modest but durable gains in living standards without compromising environmental security. New forms of these are constantly being devised.

Concluding remarks

Having reviewed three examples of national vulnerability to natural disasters in Asia, we may now evaluate the similarities and differences between them. Average annual death tolls in disasters are about 12 700 in China, 2100 in the Philippines, and 40 500 in Bangladesh. Mean numbers of people affected are

3. Perhaps the Grameen Bank system, which socializes savings and loans among poor women, may be regarded as this, although in reality this spectacularly successful initiative is a home-grown innovation.

respectively 23.4 million, 1.9 million and 11.4 million, although this is 2.02 per cent of the Chinese population, 3.08 per cent of the Philippines total and 9.5 per cent of the Bangladesh population. In this sense, Bangladesh is the worst-affected country. But in terms of the shear scale, frequency and distribution of natural disasters, China is nevertheless a worthy candidate for the title of the world's most seriously impacted nation.[4] At the same time, the scale of the environmental problem in the Philippines, and the way it interacts with natural catastrophe and poverty, highlight the seriousness of that country's plight.

The key issues are population, poverty and equity, and all three countries have heavy burdens to bear in this respect. If these factors intensify socio-economic polarization, they will have a negative effect on resistance to disasters in all three countries. Although the degree of dependence on outside aid varies considerably among the three, developments in the world economy will have profound effects, both direct and indirect, in all cases. Hence, it remains to be seen whether gains in civil protection and environmentally sustainable growth will be offset by new sources of deprivation and vulnerability. Rising population brings other problems in its wake, especially in terms of pressure on environment and resources. It is not reasonable to regard that pressure in crude terms of excessive demographic growth among the poor, for it is much more a question of the way in which resources are apportioned. But resource allocation is the key to making life safer against natural disasters, although it is likely to be a thorny issue in the future for China, the Philippines and Bangladesh.

Let us now turn our attention once again from the broad picture of national vulnerability to a more localized scale and examine the nature of post-disaster relief in a single developing country.

Yemen floods

An illustration of relatively modest relief effort is provided by the floods that occurred in the Yemen over 14–16 June 1996. Heavy rain fell on four of the country's governorates, which had a total population of 1.9 million, about one third of which was urbanized. Some 338 people were killed and 22 842 families were left homeless by the floods. A total of 43 health centres and 53 schools were damaged or destroyed.

---

4. In 1997 3200 Chinese were killed by natural disasters, 2.87 million homes were destroyed, 50 million ha of crops were damaged, and 478 million people (40 per cent of the population) were affected. In that year, natural disasters cost China US$24 billion, and of this US$530 million was spent on disaster relief.

Over the ensuing two months, donations to the Yemen's international relief appeal arrived from 23 countries and three groups of supernational organizations: United Nations agencies, the European Union, and the International Federation of Red Cross and Red Crescent Societies. The total cash value of donations amounted to US$10 276 450, as logged by their coordinator, the UN Department of Humanitarian Affairs,[5] based in Geneva. These funds were used for the purchase of drugs, food, blankets, tents, water supply and purification equipment, and health kits. They were also used for emergency repairs and to fund the missions of experts, such as water engineers. Only 1.4 per cent of the funds came from private sources and hence the vast majority represented government sponsorship, which was however partly disbursed through NGOs.

As one might expect, nearly 70 per cent of donations came from within the Middle East; 18 per cent came from Europe, 9 per cent from the supernational agencies, and the remaining 3 per cent from North America and the Far East (Fig. 8.5). But the level of donation within the Middle East region was very varied. For example, almost half of the total donations came from Syria and about 10 per cent from Qatar. Even allowing for possible inaccuracies in the UNDHA's (or UN-OCHA's) data, this implies the adoption of varied criteria of donation, especially as the sums involved do not correlate with the individual countries' GNPs.[6] In such cases, strategic alliances usually influence the decision to be open handed or restrictive with natural disaster funds.

The Yemen has been persistently troubled by factional fighting, which has complicated the relief of natural disaster. However, few countries have suffered so greatly from the twin hazards of natural disaster and warfare as Somalia, as will now be shown.

Somalia

In Somalia, climatic disasters have combined with military anarchy to create a situation in which humanitarian needs have remained consistently high and unfulfilled. It is a good example of the worst that can occur in the wake of colonial arbitrariness, superpower rivalry and militarized ethnicity, against a background of catastrophic droughts and floods (IFRCRCS 1994: ch. 7).

The 637 657 km^2 of Somali territory are draped around the Horn of Africa and contain approximately 6.76 million inhabitants at the low average density of

---

5. Now the Office for the Coordination of Humanitarian Affairs.
6. The rationale is that the Rio convention UNCED obliges wealthy signatory countries to devote a given minimum proportion (0.7 per cent) of their GNP to international development aid.

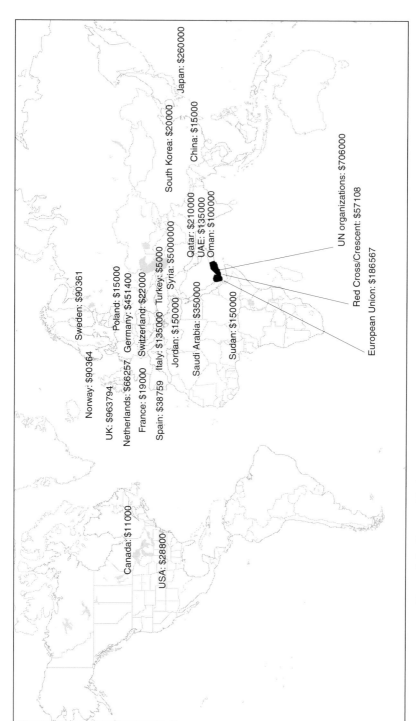

Figure 8.5 Geographical distribution of international donations to the 1996 Yemen floods appeal.

11 per km². Although there are some minerals (notably iron, uranium, and limestone for cement), life in Somalia is overwhelmingly agrarian and is based on the cultivation of cereals, cotton, sugar cane and bananas, and on animal husbandry and coastal fishing. Long dry seasons are interspersed by a short rainy period in January (the time of the *dehr* harvest) and a longer period of rain in August, when there is the *gu* harvest. Irrigated farming takes place between the Juba and Shebelle Rivers south of Mogadishu, but drought is a widespread and persistent problem, for Somalia is a country of water shortage.

Over the period of 1968–92, disaster killed an average of 21 700 Somalis per year and affected a further 334 000. Drought-induced famine was commonly at the root of the problem. In 1987, the last relatively stable year in the country, average life expectancy was 47 years, infant mortality was high at 150 per 1000 live births, and mortality of children under 5 years of age was 190 per 1000. No official statistics were collected in subsequent years, but the health and nutrition situation worsened dramatically in the early 1990s and mortality rose steeply (Hitchcock & Hussain 1987, Bradbury 1998).

The preconditions for strife could be found in the ethnic tensions and authoritarian government of the 1980s. Civil war began in 1988 and soon enveloped the country. In January 1991 President Mohammed Siad Barre was overthrown and anarchy prevailed. The country was left with US$1.9 billion of national debt, which it had no means of repaying, not least because government, taxation and official institutions had all ceased to exist. The north of the country became the self-proclaimed state of Somaliland, with 1 million inhabitants, but it has not been recognized diplomatically nor established as a working entity. Somalia proper is dominated by 15 militias, based on clans and kinship groups, which mostly use weapons that were looted from the Somali Army, to whom they were originally supplied by the USA and USSR during the superpower rivalry of the Cold War. The result has been fierce factional fighting, especially between the Somali National Alliance, under its leader Mohammed Farah Aideed (succeeded after his death by his son), and the Somali Salvation Alliance (under Ali Mohammed Mahdi). In rural areas there is strong insurgence of the Rhanweyn Resistance Army guerillas. In Mogadishu, factional conflict had become so strong by mid-1996 that most humanitarian aid work had to be suspended. Overall in Somalia, infrastructure and services collapsed in places where the fighting was heaviest and most persistent, although unofficial schools and clinics have managed to survive somewhat miraculously in many localities (IFRCRCS 1994: ch. 7).

At the height of the civil war, more than 800 000 Somalis became international refugees, while more than 1 million were displaced internally, mainly to urban areas (Ryle 1992). One contrasting example is that of Hargeisa in the northwest of the country, where in 1988 government bombing raids caused great destruction. Over the next three years the entire population of 600 000 was forced to leave. The city has traditionally suffered from chronic water shortages, and its recovery after the civil war was slow as aid agencies and technicians struggled to provide sources of clean and bacteriologically pure water for the returning population (IFRCRCS 1994: 89–91).

Inability to cultivate areas subject to fighting, intimidation, conscription, drought and traditionally low yields of crops has meant that Somalia has suffered a constant food deficiency of up to half of annual needs. By mid-1992, 4.6 million Somalis were in need of food aid and 1.5 million were at severe risk of starvation. In 1991 and 1992, disease killed between 300 000 and 500 000 people, most of whom were children and infants (Moore et al. 1993).

US-led humanitarian and peacekeeping operations with United Nations backing began in August 1992 and ended in March 1994, and multinational forces (UNOSOM) lasted from 1993 until March 1995. UN-sponsored peacekeeping is estimated to have cost ten times the value of all humanitarian aid. The Red Cross commented, "As always, any confusion between political, military and humanitarian roles, motives and actions hampers efforts to meet the needs of the most vulnerable" (IFRCRCS 1994: ch. 7). Although US–UN intervention was ultimately unsuccessful in bringing peace to Somalia, it did make considerable progress in reducing disease and malnutrition. Nevertheless, in mid-1996 more than 1 million Somalis remained dependent on food aid, including 600 000 who had recently returned to their homes, 250 000 who remained internally displaced, and 470 000 international refugees distributed among Djibouti, Eritrea, Ethiopia, Kenya and Yemen. The Lower Juba region was at this time the location of a cholera epidemic.

On 5 November 1995 intense rains and flooding caused extensive damage to parts of Mogadishu; buildings collapsed, 20 people were killed, 17 000 families were affected and many wells were contaminated. The 54 000 people displaced by the floods were accommodated in eight camps situated around the city. Seven months later, 57 000 people were affected by flooding along the Lower Juba River, which rises in the central Ethiopian highlands and which had suffered unusually heavy rains. Two years of severe drought had led farmers in the Juba floodplain to break through dykes and levees in order to irrigate their crops, and the breaches worsened the effects of the floods. This represented a typical meteo-climatic situation, in which years of drought were interspersed with occasional runaway floods. Fortunately, the latter offer some chance to produce crops if sufficient seedstock can be distributed to take advantage quickly of the recession of the waters.

At the end of 1993, 60 non-governmental organizations were at work in Mogadishu alone, but the deteriorating security situation led 54 of them to pull out (leaving only 30 NGOs in the whole country by the end of 1995). The common and repeated problems that aid agencies faced can be summarized as follows:

- theft of food, medicines, fuel and vehicles from warehouses and convoys
- aerial and ground attacks on relief supplies; arson and destruction of aid materials
- attacks and ambushes on food delivery convoys
- attempts to impose tolls and taxes on the movement of humanitarian aid
- intimidation, kidnap and murder of relief officials
- use of humanitarian aid for provisioning soldiers and guerillas
- shelling of hospitals and their expropriation to become military bases
- manipulation of agricultural and refugee situations for military purposes
- looting, extortion and banditry in rural areas
- arbitrary restrictions on which agencies could work in Somalia and where they could operate
- closure of NGO offices and confiscation of property by soldiers.

The net result has been "donor fatigue", a reluctance on the part of the richer countries to continue supplying aid and relief. So persistent is the fighting and anarchy that donors are not seeing very significant net benefits in projected future involvement. But at the same time, humanitarian needs have not diminished for years, and malnutrition and cholera remain persistent problems.

At the root of the problem in Somalia is the legacy of colonial partition in Africa and of superpower competition for hegemony during the Cold War. The powers that created the problem lack the will, the degree of engagement and the apparent means to solve it. Foreign involvement in Somalia has been inconsistent. It has responded as much to domestic and international concerns outside the country as it has to the needs of the Somali people. The result has been an erratic and unreliable supply of aid. In the United Nations' efforts in Somalia, there has been a disparity between peacekeeping, which was largely unsuccessful, and humanitarian assistance, which was largely successful, yet the one was essential to the other. The failure of peacekeeping reflected diplomatic inability to reach agreements between the warring factions and the apparent impossibility of subduing them militarily. Yet the key to the problem may well be reform of the UN structure and the proposal to set up a permanent multinational peacekeeping force to be deployed in the world's trouble spots. Much depends on whether such initiatives ever come to fruition.

The natural disasters that take place periodically in the fragile desertifying landscape of Somalia are superimposed on the human tragedy of conflict, and they aggravate the situation. In many places it is impossible to begin to construct new means of reducing vulnerability to the creeping disaster of drought and the sudden one of flood. The government structure and national infrastructure, the necessary peace and stability, are no longer there.

Various lessons can be learned from disasters in the 1990s in Asia, Yemen and Somalia. To begin with, most of the situations that have given rise to casualties and losses have been subtle and complex: the impact of natural and social disaster is only one aspect of a situation that has usually taken a long time, perhaps even decades, to develop to crisis proportions. In many cases natural disasters are merely punctuating events in a constant stream of misfortunes: normality is a disaster, peace and security are seemingly unattainable goals.[7]

In these "complex disasters" (Duffield 1996) it has proved harder to tackle the causes than the consequences, yet the solution lies in the former, not the latter. As never before, international relations have dominated the scene. The root of the disaster problem is a mixture of military or political opportunism, diplomatic self-interest and, in third place, natural extremes. Motives are hard to analyze, but the international aid agencies have struggled to maintain their neutrality in this difficult situation, although many now question what neutrality means in the face of repression, aggression, manipulation and violation (Slim 1997).

The next section will present a portrait of one such agency, which operates under United Nations auspices, in order to show how difficult disaster relief has become in the world's main trouble spots. To the individual victims it matters very little whether the disaster is natural, social or technological in origin. What is needed, and what is becoming increasingly difficult to provide, is comprehensive relief, security and stability.

World Food Programme operations

A detailed picture of the relationship between security and disaster relief is furnished by the following account of the UN's World Food Programme operations over the four months from mid-November 1995 to mid-March 1996. The period has been chosen quite arbitrarily and is entirely characteristic in terms of the sorts of problems that the WFP has to cope with in order to supply food to those in need. During the four months in question, the WFP concentrated its operations in ten different areas of the world (Table 8.1) and worked closely with other UN agencies and with NGOs.[8]

As one might expect from the state of the world, the principal problems involved military conflicts, insurgence and sporadic violence, although not

7. These are situations that Rudyard Kipling called "The savage wars of peace". See Macrae et al. (1997) for an example.
8. Data for this section were obtained from the World Food Programme weekly reports supplied by Volunteers in Technical Assistance (http://www.vita.org).

Table 8.1 Use of World Food Programme resources, 1996.

Country	People needing aid (millions)	Needs	Shortfall	Shortfall (%)
Angola	10.6	90720	66570	73.4
Ethiopia	4.2	76000	52670	69.3
Georgia	3.0	25733	19039	74.0
Iraq	21.6	301051	239085	79.4
North Korea	5.0	15785	9285	58.8
Rwanda	24.5	500608	239065	47.8
Sudan	2.0	20246	8000	39.5
Tajikistan	4.0	32850	24624	75.0
Uganda	2.5	54625	29625	54.2
Ex-Yugoslavia	22.6	330657	198948	30.1

outright warfare. Yet at the same time a significant number of natural disasters occurred, and hence the overall situation in each region reflected the merging of natural and socio-military disaster in such a way that the two entities could hardly be distinguished from one another, at least in terms of the operational problem of providing food relief to the hungry. The only WFP operation directed solely to natural disaster conditions was that in North Korea (the Democratic People's Republic of Korea), where floods in early December 1995 and concurrent hailstorm damage to standing crops led to famine conditions in several provinces. In mid-December, UN operations in Bosnia were hampered by heavy snowfalls. But the principal natural disaster was drought, which was widespread throughout six countries of southern Africa. It affected 945600 people in Mozambique, 350000 in Lesotho, and 75000 in Swaziland, where it had lasted four years. Drought also occurred in Malaysia, Tanzania and Zambia. In January 1996 it led to crop failure in Angola and Rwanda, whence it spread to Burundi. Water-table drawdown and accelerated soil erosion had serious impacts upon Swaziland, to the extent that drought had deepened into desertification in some places. Drought in Tanzania required 80000 tonnes of food aid to be distributed to 550000 recipients, and at the same time an invasion of red locusts cost almost US$1 million to contain.

Against this background of natural hazard and disaster, conflict erupted again and again, with severe repercussions for humanitarian aid. Food delivery convoys were repeatedly harassed by militias in Bosnia and Somalia, where a driver was injured by gunfire. The vehicle of a UNHCR field representative was hijacked in Tanzania in February 1996. In the same month, attacks, gunfire and mortar bombings disrupted food distribution programs in Sierra Leone and Liberia. Curfews were imposed at night time in various cities in the affected countries.

During the four-month period under examination, by far the most serious problems were experienced in Zaire (now the Democratic Republic of the Congo), Rwanda and Burundi, where the security situation was both complex and precarious. In essence, the conflicts in Rwanda and Burundi spilled over into Zaire. Rwanda, in particular, supplied many refugees to both Zaire and Tanzania. In the latter country some 344 000 Rwandans were accommodated in the camps situated 15–30 km from the border. Violence occurred repeatedly in the Goma and Bukam Districts of Zaire. In part this was because of tensions imported from the neighbouring countries, but it was as much the fault of Zairian forces. The gendarmerie had aligned itself with the Hutu faction, while the "Red Berets" of the regular Zairian army aligned with the rival Punde ethnic group. The WFP described the Red Berets as having no chain of command, no discipline and the tendency to act individually and with violence. Thus, in Goma, 25 December 1995 was marked by gunfire and looting by Zairian soldiers. The year ended with another feud and a major explosion. Next month some of the refugees arriving in Zaire from Burundi had gunshot wounds. At Bujumbura in Burundi all UN activities had to be suspended for three days as a result of heavy machine-gun fire. In fact, gunfire occurred nightly. Grenade, gunfire and machine attacks in Burundi left at least 176 dead at seven separate locations on 6 December. Also in Burundi, a grenade was thrown into a hospital ward, causing injuries. In February in Rwanda unidentified gunmen blew up a water pipeline and an electricity plant in Goma, where a nocturnal curfew was necessary. At Cyangugu in Rwanda, shootings, grenade explosions and other incidents took many lives. At Kahindo refugee camp in Zaire, two children were shot dead in January 1996 by the Zairian Army contingent of camp security, apparently without motive.

UN and NGO vehicles provided easy targets for soldiers and armed bandits. In Bukam, Zaire, an Italian driver was shot in the attempted theft of his vehicle. In December a Belgian vehicle was assaulted in Burundi and workers travelling in it were robbed and their military escort was killed. Three aid vehicles were stolen at gunpoint in Burundi in November and eight were stolen in December in Goma. In Rwanda two vehicles were stolen and their custodian was seriously injured. Vehicles were susceptible to damage by landmine. A UNHCR water truck was badly damaged by a landmine on a section of Zaire road where an aid worker had previously been severely injured when he stepped on a mine. Five Zairians were killed and ten injured in Bukam when their minibus ran over a mine. In Rwanda a vehicle belonging to the International Rescue Committee was destroyed by an anti-tank mine in February. Explosions and further damage to the WFP trucks led to restrictions being imposed on food distribution as a result of landmine hazards.

But the landmine problem was not restricted to transportation. Landmine explosions were heard sporadically each night in Goma. Refugees from Rwanda were killed and injured by mines in January. But when roadside mines killed five Zairians in Bukam, the local population turned on the Rwandan refugees and beat three of them to death. Moreover, the Zairian Army threatened to deploy 500 mines in order to discourage the movement of refugees into and out of Kibumba camp. At the same time, the Burundian Army had already mined the country's border with Zaire in order to restrict the flow of refugees to the crossing points that it manned. This had the joint effect of restricting the influx of refugees into Unva (Zaire) to 160 per day and ensuring that many had mine-inflicted wounds.

Shootings, grenade attacks, intimidation, death threats, vehicle hijacking, reprisals, curfews and the ubiquitous landmine problem combined to restrict the ability of the WFP and other aid agencies to supply food to the civilian population, whose ability to produce and market its own food had been reduced by conflict and drought. Food shortfalls in the Zaire–Rwanda–Burundi operations for the six months ending in May 1996 were estimated at 58 072 tonnes of cereals, 12 184 tonnes of pulses, 3061 tonnes of oil, and 4440 tonnes of other foodstuffs, a total of 77 758 tonnes. In November 1995 2 440 000 people needed to be supported and only 4815 refugees were repatriated. As an example of how long-term estimation could not be matched by long-term planning, food stocks in Rwanda were estimated in 31 November 1995 to be sufficient to meet 26 days' requirement in cereals, 24 days in pulses, and 39 days in oil. At the same time, a government review showed that, of 158 NGOs operating in Rwanda, 38 had failed to register their activities officially or did not meet adequate standards and would therefore be asked to leave.

In the ten regions where WFP activity was concentrated at the end of 1995, five involved protracted refugee situations (Angola, Ethiopia, Sudan, Uganda and the former Yugoslavia), three involved large numbers of internally displaced people (Angola, Georgia and Tajikistan), three involved the effects of war (Angola, Rwanda and Tajikistan), one was the result of floods (North Korea) and one of general vulnerability (Iraq). Almost 10 million people required food assistance during 1996 and an estimated total of 1 117 618 tonnes of food was required. In February 1996 the shortfall (1996 requirements, minus available stocks and contributions promised) amounted to 6.32 per cent.

The provision of food aid is thus rendered difficult by expediency, or inability to make provision very far in advance, and by uncertainty. The latter has many facets. Will donors come up with the necessary aid in time and in quantity to be able to ship it to those in need? Will further sources of need emerge and thus inflate the food requirements? Will international politics reduce the scale of giving? Will military situations on the ground worsen or will natural disasters occur and will local food production suffer further adverse impacts? These are indeed precarious times.

It is perhaps not that disasters have become more readily complicated by politics in recent times, but rather than relief has, and also that politics themselves have, become more complex with the loss of simple power-bloc polarization at the end of the Cold War. Complex emergencies are characterized by the need to provide relief to civilian populations in various stages of destitution as a result of natural disaster, environmental despoliation, military offensives, low-level guerrilla action, or political and socio-economic breakdown. Generally, these elements occur in some combination that reinforces the misery of the affected population and increases the number of casualties. In the worst cases (for instance, Somalia in the early 1990s and Sudan in the late 1990s; Bradbury 1998), natural disasters such as drought and flood hardly alter a situation that is already catastrophic for military and economic reasons. For example, the Afghan earthquake of February 1998 killed 4000 people but did not stop the factional fighting in the civil war that for years engulfed the northern provinces of Afghanistan.

As a result, it is often difficult or impossible adequately to distinguish the phases of natural disaster from the generally precipitous situation. Under stable conditions, impact is followed by gradual remedy; in complex disasters, breakdown in relief efforts or failure to cope with renewed military threats may lead to a worse disaster than the initial one that stimulated the relief action in the first place, and the seriousness of the problem may oscillate rather than ameliorate. Moreover, there is a growing realization in the world relief community that neutrality is difficult or impossible to maintain under the duress of repeated attempts to subvert aid and relief for military purposes (Hendrickson 1998, Leader 1998, Macrae 1998). The only answer is to be able to restore peace, economic stability and a modicum of democracy in order to integrate disaster prevention with general economic development. Under the circumstances this is the only acceptable form of symbiosis. However, it is often doubtful whether such goals can be achieved while still preserving the traditional neutrality that is the only guarantee that relief will be given impartially to those in need.

These problems have given rise to a considerable debate among the relief community and have led to a renewed consensus on the need to link disaster mitigation to economic development and environmental protection.[9] Yet there have been dissenting voices, which have argued, for example, that all emergencies are complex and hence the epithet "complex emergency" is redundant (Kelly 1996). However, there is little doubt that the problems of marginalization and extreme vulnerability to disaster result not from single problems but from a plexus of threats, in which the military and humanitarian aspects of aid interlink in complex ways. A simple diagrammatic explanation of how this occurs is given in Figure 8.6.

---

9. The debate is well summarized by Duffield (1996) and IFRCRCS (1997: ch. 2).

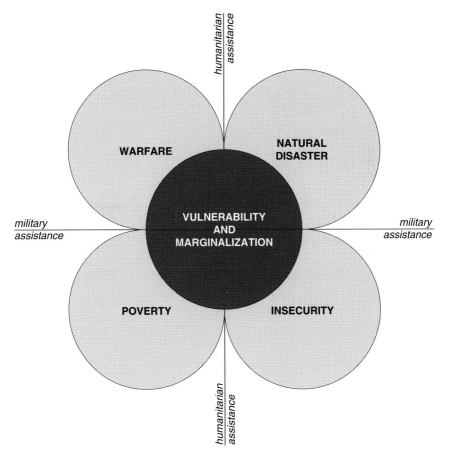

Figure 8.6 The "military cross": interlinking of military and humanitarian assistance in the creation or mitigation of disaster.

A local perspective

We will now consider two examples from the Peruvian Andes to illustrate aspects of disaster in the context of the low-level emergency that is daily life. Natural disasters such as floods, earthquakes and landslides should thus be considered as "emergencies within emergencies" in such a context.

All for a sack of potatoes

Some years ago I witnessed a scene, as comical as it was sinister, that highlighted the way in which social tensions can exacerbate vulnerability to disaster. Disaster in this case might equally have meant earthquake, flood, landslide, terrorist attack, military repression, or warfare, or any combination of these. The incident took place in a small, fairly impoverished Andean town, a place of grey trachyte[10] walls, clear but rarefied atmosphere, snow-capped mountain horizons and ubiquitous dust. In the midst of one of the main thoroughfares a street market was in progress. The usual mixture of local produce, snacks, hardware, livestock and tourist gewgaws were on sale. Latin music blared out of tinny loudspeakers and there was a general air of animated tranquillity.

Suddenly an olive-coloured truck drew up and a detachment of Peruvian army soldiers disembarked to buy a sack of potatoes. The driver remained in the vehicle with the engine running and a machine gunner to guard him. Four soldiers took up firing positions at the corners of the truck and four others concealed themselves behind walls a few paces away, dropped to one knee and raised their rifles. The remaining two bought the potatoes and hauled them to the truck. Their colleagues then retreated in orderly fashion and leapt aboard the vehicle, which, bristling with rifles, set off in a cloud of dust and exhaust fumes. None of the market traders or their customers paid the slightest attention to this performance.

Some years previously the town had been seriously damaged by earthquake; in fact, reconstruction was still not complete. Since then it had been constantly under the threat of guerilla attack, a reality held at bay only by the presence of sizeable garrison of army conscripts. Given the paucity of civil protection resources and structures, the army is destined to play a key role when natural disaster strikes such a town.[11] Yet the reality of the situation is that the military and political tensions would be likely to restrict severely the army's freedom of operation and to colour strongly its relationship with the civilian population of victims and survivors. No more powerful argument can be made for developing a contextual model of natural disaster.

A second Peruvian example will serve to illustrate the holistic nature of disasters that affect marginalized communities in developing countries and in particular will show the problem of vulnerability as it is viewed by those who bear its burden.

---

10. A light welded rock, or tuff, produced by highly explosive volcanic eruptions.
11. In the El Niño floods of 1997–8 the relationship between the central government of President Alberto Fujimori, the Peruvian Army, local government relief committees and the people became decidedly complex and acerbic.

Floods and landslides in the Andes

The Southern Peruvian town of Cuyocuyo (pop. 5500, altitude 3500 m above sea level) lies in a deep valley among the grey slate mountains of the Andean Eastern Cordillera. The bare debris-mantled slopes rise steeply 1000 m to a rocky shoulder, above which snow-capped peaks tower a further 600 m high. Above the valley a rolling altiplano is dotted with grazing llamas and alpacas, but there are too many of them and the hardy brown grass is thinly distributed and riven by muddy trails, a situation amounting to desertification.

Cuyocuyo (Fig. 8.7) is a poor and isolated community of farmers who till the banks of stone-fronted terraces that line the valley sides in spectacular flights. These were originally constructed by the Huari peoples in the thirteenth century and many are abandoned. Rockslides and falls are frequent and destructive, and hence many of the terraces have collapsed into long cascades of stones and earth. It is back-breaking work to rebuild them. Neatly hoed, the thin dusty upland soils of the terrace benches yield modest crops of small potatoes, tomatoes and onions. On the tiny verges between the crops, draft animals mingle with lactating alpacas with offspring attached.

The town winds along the banks of the Rio Cuyocuyo, which trickles down from the brilliant white snow-capped peaks (*nevados*) in the distance. The houses are mostly built with thick drystone walls of reddish-grey slate and have

Figure 8.7 The town of Cuyocuyo in the Eastern Cordillera, Department of Puno, southern Peru.

221

roofs of thatch or *lamina*, the ubiquitous sheets of corrugated steel. They are largely unadorned. No doctors, nurses or policemen live in Cuyocuyo. There is no electricity or gas, and very little wood. Only kerosene is used as a fuel, to light the lamps and cook meals. Goods are brought in for sale along the winding dirt road across the spectacular scenery of the Cordillera. The aging trucks, with their tall, slatted wooden sides, sway and shudder as they make the 225 km journey from Lake Titicaca, heavily laden with people, food, fuel and livestock. Sixteen hours after leaving Juliaca, the people of Cuyocuyo alight in their village, covered from head to toe in a pall of reddish-grey dust. And every so often a truck goes over the edge of the steep and rutted track, tumbling down the rock slope into the dark ravine beneath. In the rarefied air of high altitudes, the sun shines intensely on Cuyocuyo, but only in the middle of the day, for until mid-morning and from mid-afternoon the sides of the valley cast a deep, opaque shadow across the angular shapes of the houses. At mid-day, currents of air waft trains of puffy white cloud into the valley, and these curl around the mountain peaks that tower above the settlements huddled in the valleys or draped over the hillsides. The people of Cuyocuyo are *indios*. The Spaniards conquered them, introduced a nominal anthropomorphic form of Catholicism and left. The *mestizos* (half-breeds) who resulted from such contact also abandoned the valley for the easier life and richer pickings on the Altiplano.

Life expectancy is short in Cuyocuyo, about 54 if one discounts infant mortality, otherwise it is 36.[12] Work is organized communally and shirkers are not tolerated. Hygiene is a very relative concept and illnesses are cured by folk remedies or are left to run their courses. Besides the sporadic passing of trucks, there is little contact with the outside world, and hence the community has remained untroubled by the terrorism that has ebbed and flowed through the other valleys of the Cordillera. Hospitality is limited by poverty and by a deep-rooted suspicion of outsiders. The Cuyocuyeños are taciturn but respectful, deliberate in their demeanour, at least until those rare occasions when alcohol flows in their veins. Their music is syncopated and primitive, but it echoes majestically, rolling across the portals of the valley like thunderclaps.

Natural disasters intrude with ease into such a setting. Eight phases of orogenesis have heaved the Cordillera out of the sea and as high as 6000 m, and so the landscape is richly strewn with the geological evidence of vast and cataclysmic events. At Sollanque on the downstream fringes of Cuyocuyo there are the scars of a massive landslide that once sent 600 million m^3 of rock hurtling into the valley beneath (Alexander 1992). Water was impounded behind it until the pressure was so great that it burst through the debris barrier and scoured its way down stream towards the Amazon headwaters in an immense flood. This

---

12. I am indebted to Dr Bruce Winterhalder for this information.

prehistoric cataclysm was followed by many other events, some of which were recent enough to damage the town.

In 1985 the rains were long and copious. Moisture bubbled up in the thin soils of the steep slopes and these slid into the deep creases of the ravines. The Rio Jilari, just up stream of Cuyocuyo, was set into motion and rolled down valley in a thick turbulent stream of water-laden debris blocks, a geomorphological midpoint between a viscous debris flow and a less dense hyperconcentrated streamflow. The morass of slate blocks tumbled into town, buried 200 houses, and demolished the parish church. The townspeople had heeded the state of the weather sufficiently to have evacuated. But an old lady and her niece had gone back to retrieve their valuables from under the floorboards of their house and they were killed and buried by the debris.

The tongue of debris and black bumpy fan of rock fragments remained as mute testimony to the harshness of the environment (Fig. 8.8). The broken walls

Figure 8.8
Flood and debris flow deposit (outlined in white) in Cuyocuyo, southern Peru.

of houses appeared at intervals among the wastes of rock. A year later, when the rains returned, water saturated the valley again and bubbled out of the debris, flooding its way across the rest of the benighted village.

Clearly, new houses and a church needed to be built and the Rio Jilari needed to be stabilized and canalized against further debris flows. The mayor of Cuyocuyo went to Lima in search of help (a journey of monumental proportions and no small expense). He toured the foreign embassies and NGOs in search of funding and expertise. But Peru is a poor country with much destitution and phenomenal environmental problems and he returned empty handed. In the next year, modest attempts were made to rebuild the church and clean out the channel of the Rio Jilari, but the northern part of town remained buried under debris.

In this simple portrait we see a remote and marginalized community afflicted by a disaster of relatively modest proportion but enduring impact. The potential solutions to the slope instability problem at Cuyocuyo are numerous, but many of the means are lacking. However, it should be borne in mind that a tectonically active valley with 42° slopes 1000 m long is bound to be unstable. The miracle is that so much has been achieved by terracing, a patient art of centuries, which has built up a subtle response to the mobility of slopes, and one that is durable if maintained.

At Cuyocuyo the community is well organized, but it has few resources and no particular means of generating more of them. Inhabitants find themselves in a poverty trap. As local economic opportunities are so meagre, the panacea for personal advancement is considered to be out-migration. This is often a one-way process that brings little benefit back to the village. Given this continual depletion of the population, one may wonder whether Andean settlement of this kind is viable at all. It is deficient in healthcare, social security, infrastructure, goods, services and market support. Government at all levels is remote and uninterested, choices are limited, and social support in the form of community solidarity is all that makes life in such places sustainable (Winterhalder & Thomas 1978).

In terms of local involvement with environmental problems, and the sophisticated degree of social organization that is present within the community, Cuyocuyo demonstrates considerable potential for natural hazard mitigation (it also faces a considerable natural hazard risk). But the lack of economic viability that is inherent in marginalization leads to stagnation and inability to reduce risks. In such an environment, natural hazards cannot be considered as separate entities. They are an integral part of the risks of being alive, which encompass a much wider range of preoccupations. Two years after the debris flow first entered town, a young bride was working one day on the agricultural terrace high above Cuyocuyo. She was surrounded by a group of about 20 of her in-laws and she had already gained a reputation for laziness and rebelliousness. Next morning her body, with the neck broken, was found at the foot of the slope beneath where she had been working. It was deemed an awful warning that the community cannot tolerate dissidence and sloth. The harshness of life in such a place does not permit such latitude.

We shall now return from the particular or local scale to the global, and draw some conclusions about disasters in developing countries. With regard to future trends, the signs are both ominous and encouraging.

The way ahead

Interest in natural disasters has never been greater than it is at present. Through desire to mitigate risk, or through the sheer necessity of providing relief, governments and NGOs have become ever more deeply involved in the fight to prevent casualties and catastrophic losses. But these efforts have become strongly polarized. The industrialized world has invested heavily in technological solutions to natural hazards (such as monitoring and warning systems and expensive structural defences), while the developing nations have had to rely on management of human resources, often against a background of complex political, military and environmental instability. There is little sign of any great degree of technology transfer, or of any great willingness on the part of the developed world to share its sophisticated monitoring and warning systems with the developing countries.

One is led to ask whether the world can sustain its current level of interest in disasters and, above all, whether it can translate that interest into better standards of mitigation? In the industrialized nations the relentless rise in the value and vulnerability of fixed capital assets will inevitably lead to future losses that are costly enough to stimulate further mitigation efforts. Surprisingly, although mitigation is invariably cost effective in terms of damage avoided, there have been very few cost–benefit analyses, and hence it is often difficult to convince governments to invest in disaster prevention measures that do not show immediate benefits.

With regard to developing nations, the 21 principal donor countries increased their humanitarian aid quotas nearly six times from 1985 to 1994 (to an annual total of US$3.47 billion; IFRCRCS 1996). In 1995 alone, the Red Cross launched 55 relief appeals (9 of them exclusively for natural disasters), which brought in US$270 million. Disaster aid is thus very big business indeed. However, the boom in international relief that followed the end of the Cold War appears to have ended, and funding has begun to diminish, despite the continuing demonstration of high levels of need (IFRCRCS 1998). Given demands for the more effective use of increasingly scarce funds, much thought is now being devoted to ways of integrating development and mitigation with disaster relief. In this, there is a need not merely for much greater levels of transfer of technology and expertise but also for a two-way learning process. The value of this was amply demonstrated by, for example, Italian interest in the September 1994 eruption of Rabaul Caldera in Papua New Guinea (McKee et al. 1984, Denatale & Pingue 1993), which provided an important analogy for a future eruption scenario in the Campi Flegrei at Pozzuoli (Bianchi et al. 1987). Examples such as this show that the key to better reduction in vulnerability to disasters lies in sharing technology and expertise and in the better management of emergencies as social phenomena that require innovative organization.

CHAPTER 9

Finem respice

Previous chapters have considered the theoretical, social, technological, moral and development issues involved in natural catastrophe. It is now time to draw some conclusions from this survey of change and trends in a world afflicted by recurrent natural disasters. Given the accelerating pace of the transformations discussed in this book, the conclusions are destined to be both provisional and incomplete: they, too, must adapt to circumstances.

This final chapter will not attempt to review or even synthesize the diverse subjects covered in the previous chapters, but will underpin them with a comprehensive theoretical model, which, it is hoped, will help to enlighten them.

The DNA of disaster

The most widely accepted view of cause and effect in disaster can be conceptualized as follows:

Extreme geophysical events
act upon
human vulnerability and risk taking
to produce
casualties and damage

Deliberate risk taking is dominant in affluent societies where there is ample opportunity for choice, and it produces foreseeable or avoidable vulnerability. On the other hand, the risk taking associated with poverty and marginalization results in inadvertent or unavoidable vulnerability and the range of practicable choices is severely restricted. However, the view is gaining ground that the physical agents of catastrophe are not the starting point in the chain of cause and effect: rather, human vulnerability and risk taking are held to be (Quarantelli et al. 1995). One way of conceptualizing this is:

Society's risk taking and vulnerability
interact with
extreme geophysical events
to produce
casualties and damage

A more extreme reorganization of the fundamental variables gives the following chain of causality:

Human risk taking and vulnerability
produce
casualties and damage
when there are
extreme geophysical events

However, the consensus is still in favour of integrating the physical and socio-economic aspects of disaster *in that order*, and hence this is the line that this section will follow, although it will make allowance for feedback mechanisms.

One justification for this approach is that physical impact is followed consecutively by human response. Although it is not a process in its own right, time is the linear backbone of disaster, whose medium of expression is geographical space. The other principal dimensions are magnitude, a measure of the size of the physical event, and intensity, a measure of the size and extent of its effects (Alexander 1991a, 1994a, 1995a).

The temporal pattern of hazards varies considerably with the type of event, but essentially there are ten different forms in which they occur, which will now be listed with examples:

- *Non-repetitive hazards* Some hazards occur only once. This is commonly the case for the large debris avalanche–slide–flow phenomena that occur in Alpine and other mountain areas and are known as *sturzströme* (Hsü 1975).[1] Non-repetitive events are also characteristic of the interaction of natural and anthropogenic hazards, as in the case of a dam burst (Ellingwood et al. 1993).
- *Remobilized or reactivated hazards* Continuous erosion (by a basal stream or by the sea at the base of a cliff), can cause a complex of multiple-regressive landslide blocks repeatedly to pass the threshold to instability (Seed & Wilson 1967).
- *Continuous hazards* Subsidence of the ground often falls into this category, especially if compaction–consolidation of saturated granular sediments occurs constantly (Chilingarian et al. 1995).

---

1. However, there are some good examples of repetition even here: for example, the disasters of 1962 and 1970 at Mount Huascarán, Peru (Plafker et al. 1971, Browning 1973).

- *Episodic hazards.* Coastal erosion is often an episodic phenomenon (Kaufman & Pilkey 1984, US National Research Council 1990).
- *Seasonal.* Drought and floods are usually conditioned by seasonal variations in weather patterns; so are hurricanes and other cyclonic storms (Kundzewicz et al. 1993). When viewed on an annual basis, coastal erosion tends to be highly seasonal, although varied geomorphic settings lend resistance to erosion an episodic character (Herbich & Haney 1982, Dolan et al. 1987).
- *Other regular spacing.* Endogenous forces within the Earth can sometimes impart regularity or pseudo-regularity to the pattern of earthquakes or volcanic eruptions. In the former case, strain builds up gradually between fault blocks until it is abruptly released in the form of seismic energy (Thatcher 1976, Chen et al. 1998), whereas in the latter case magmatic pressure accumulates underground until it is sufficient to force its way to the surface and start an eruption (Ryan 1990). Regular volcanicity is a characteristic of so-called closed-system volcanoes, which are fed by a relatively isolated magmatic convection system (Lirer et al. 1987).
- *Irregular.* Although there is a standard frequency distribution for tsunami occurrence in the Pacific basin, the actual pattern shows a fair degree of irregularity, as not all large offshore or littoral earthquakes are tsunamigenic (Lockridge 1988).
- *Progressive with thresholds.* Accelerated erosion is often identifiable by the creation of dense systems of rills and gullies, or deep flat-floor canyons. In such cases, the erosional system has crossed a threshold from superficial slope wash with low sediment load, to the wholesale export of sediment by a process of vigorous downcutting (Cooke & Reeves 1975). Likewise, desertification involves thresholds in biophysical productivity and land capability (Phillips 1993).
- *Progressive without thresholds.* When drought extends across several seasons, its effects on biological productivity tend to deepen in a smoothly progressive manner (Kassas et al. 1990).
- *Compound or complex.* Any combination of hazards – seismic landsliding, for instance – can give rise to complex sequences. Large and prolonged volcanic eruptions are also temporarily complex. For example, as described in previous sections, Mount Vesuvius was active in several hundred eruptive phases between 1631 and 1904.

The repetitiveness of hazards can also be considered in terms of frequency distributions. At the most basic level, the magnitude–frequency principle leads us to expect many small insignificant events and, in the long term, increasingly fewer events as magnitude rises: thus, there is a negative linear-logarithmic relationship between magnitude and return period or recurrence interval (Fig. 9.1a;

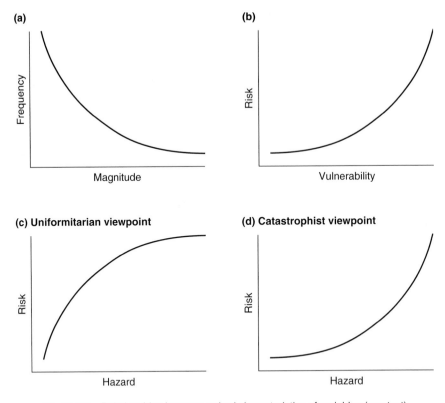

(a)

Frequency

Magnitude

(b)

Risk

Vulnerability

(c) Uniformitarian viewpoint

Risk

Hazard

(d) Catastrophist viewpoint

Risk

Hazard

Figure 9.1 Relationships between paired characteristics of variables (see text).

Alexander 1995a). To render the magnitude–frequency rule operational, we must define an arbitrary level of magnitude that represents the threshold of disaster, below which magnitudes are too small and statistically too frequent to matter (Fig. 9.2). This is justifiable because, although there are a few cases in which minor hazards give rise to major losses for any given area, there is normally a rough-and-ready equivalence between the physical event and the human scale of its consequences.[2]

The uniformitarian view of hazards translates the product of frequency and magnitude into a series of energy or impact levels that peak at a relatively low level of magnitude and high level of frequency (Fig. 9.3a). Very large events are

---

2. At times the relationship is very rough and ready. For example, good anti-seismic construction and retrofitting can prevent a shallow-focus earthquake of magnitude, say, 5.6 from becoming a major catastrophe in a nearby city. However, we should beware of assessing the strength of earthquakes or any other geophysical agents of disaster by a single measure such as seismic magnitude. The physical forces are more complex than magnitude would suggest and the disaster potential is yet more complex still (Foster 1976, FEMA 1988).

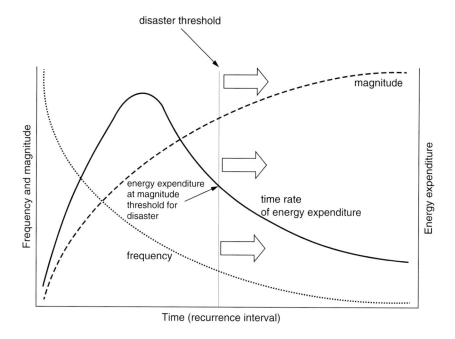

Figure 9.2 Distributions of frequency and magnitude of events in relation to the disaster threshold.

thus geophysical curiosities but are not usually fundamental to the working of the system. On the other hand, a catastrophist viewpoint translates the magnitude–frequency rule into a curve of increasing energy or impact level and thus apportions more weight to high-magnitude low-frequency events than to the combined effect of events of moderate magnitude and comparatively high frequency (Fig. 9.3b).

A similar relationship exists for risk levels in relation to the magnitude of

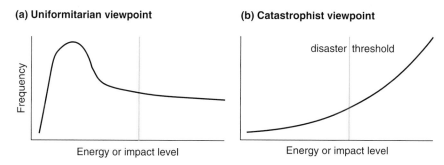

Figure 9.3 Two views of the arbitrary disaster threshold in relation to the energy or impact level of natural hazards.

geophysical hazard: the rising curve of risk can be held to level off with increasing hazard if one adopts a uniformitarian perspective (Fig. 9.1c) or to steepen if one prefers a catastrophist view (Fig. 9.1d). Yet, for a constant level of physical hazard, risk levels tend to rise steeply and disproportionately, with increasing vulnerability (in other words, high levels of vulnerability lead to very catastrophic losses, however these are measured; Fig. 9.1b). This relationship can be compared to the magnitude–frequency curve in order to derive a measure of the effects of disaster. Hence, let us define an initial impact threshold at the point where the falling magnitude–frequency curve intersects the rising risk–vulnerability relationship (Fig. 9.4). The arbitrarily defined disaster threshold will be set somewhere to the right of this. The curve $R/V-F/M$ defines the rising impact of disaster based on both physical processes and socio-economic or human responses (losses). With increasing steepness it ascends from zero (insignificant losses) to the disaster threshold (the beginning of catastrophic losses) and beyond.

If magnitude is closely related to impact level, then the cumulative effect of losses (the sum of impacts) is likely to rise with the increasing frequency of hazards. When compared to the ascending curve of risk in relation to vulnerability, this generates a series of potential relationships. If the risk–vulnerability

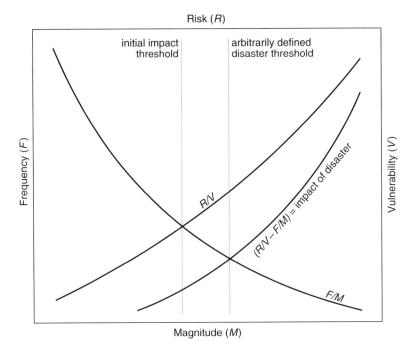

Figure 9.4 Definition of disaster magnitude threshold.

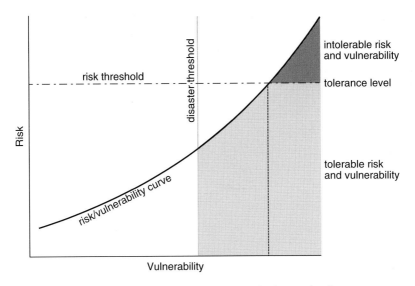

Figure 9.5   The disaster threshold and tolerance level.

relationship is significantly greater than the frequency–impact one, then considerable scope exists for hazard mitigation. On the other hand, if it is significantly less, then progress has been made in the mitigation of risks and impacts (Fig. 9.5). Whether this progress is sufficient depends on a tolerance level that is set arbitrarily and determined by societal values and priorities.

As the curve of risk against vulnerability rises, it encounters both the disaster threshold and the tolerance threshold. These define two areas beneath the curve: in the first, both risk and vulnerability are considered intolerable and are thus in need of reduction. In the second, risk is considered (by societal consensus) to be tolerable, but that still leaves a substantial area of vulnerability that could be mitigated (Fig. 9.6).

This begs the twin questions of what level of risk (or vulnerability) is tolerable and what measures can be used to reduce it? In reality, society's tolerance levels tend to fluctuate with the occurrence of disaster impacts. A serious catastrophe that causes widespread casualties and losses will create a groundswell of opinion in favour of renewed mitigation effects. It will thus reduce the tolerance level. But a long period of quiescence may allow other priorities to supplant hazard mitigation and disaster preparedness, thus increasing it.

The risks and hazards associated with natural disaster are mitigated in various ways. For new construction or new activities, sites can be selected that are the least susceptible to particular hazards or that enjoy some natural or artificial protection against impact. New structural work can be designed and executed in such a way as to be hazard resistant (flood proof, anti-seismic and so on).

233

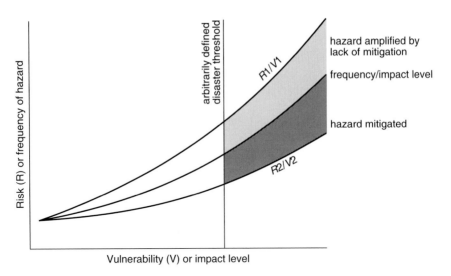

Figure 9.6 The role of mitigation in determining impact level.

Existing structures can be upgraded or retrofitted to improve their resistance to hazards. Non-structural measures associated with land-use planning, insurance, forecasting and warning can be implemented. Finally, where disaster is inevitable, response capability can be improved so as to facilitate the search and rescue of victims, repair of damaged facilities, management of emergencies and eventual reconstruction of damaged buildings. Each of these prescriptions represents a dynamic process, which, if neglected, can contribute to vulnerability. Thus, safety is not merely gained by improved organization and hard work, *it is lost by neglect*. For instance, dilapidated buildings are generally less resistant to hazards than are well maintained ones (Barbat 1996).

It should now be clear that there is a plexus of interlocking processes and responses that link hazard, vulnerability, risk and loss. Although it is not inevitably so, losses do arise consistently from both the interaction and the summation of hazard, vulnerability and risk. A pessimist would see a certain level of temporal determinism in this: given the state of the world (or so it would be argued), vulnerability must inevitably worsen, risk levels must inevitably rise, the threat of disaster must inevitably transform itself into impact, and, as time's arrow pursues its linear course, disasters must be transformed into the inescapable impress of history upon culture.

This represents a metastable system with a positive but eventually unsustainable trend. Vulnerability and losses rise relentlessly but are periodically held in check when disaster occurs and public opinion demands a sudden effort at mitigation (see Fig. 2.4). Yet the socio-economic forces that impel them to rise – population pressure, unprotected urban development, changing human expectations, and so on – soon restore the upward trend. Of course, this is a simplification of complex processes. At the same time, knowledge and expertise on disasters and how to mitigate them are accumulating rapidly. New initiatives are gathering force and will one day have a significant impact on losses, although the solution to the disasters problem is evidently not a short-term one. Yet the trend is so patently unsustainable that eventually some scale of human or economic losses will provoke a strong enough counter-reaction for mitigation to be successful. It is doubtful whether any single year will eventually appear as a turning point; it is instead likely that present models of population growth, economic expansion and social expectations will endure until at least AD 2050. When the forces that exacerbate vulnerability to disasters have significantly changed, we will be able to identify a transition to sustainable mitigation.

In synthesis, disasters are the product of a series of opposites or balanced factors that create tension. Time and space represent the first pair and the fundamental media of catastrophe. Thereafter, chance vies with inevitability in the unfolding of events. Among the responses to these, mitigation vies with amplification, and technology with culture, in the way that solutions are devised. We can view this situation in two ways. First, it might be regarded as a wheel of fortune in which there is a ceaselessly revolving interaction between the core concepts – vulnerability, risk, hazard and losses, and the pairs of processes and responses (Fig. 9.7). A second way of looking at the matter is to consider each pair of variables as part of a plexus intertwined around the fundamental skein

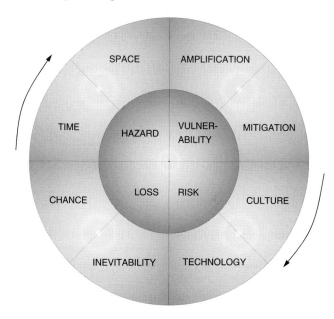

Figure 9.7 The natural disasters "wheel of fortune".

of time and space (Fig. 9.8). Thus, we arrive by analogy at the "DNA of disaster", a spiral of positive and negative attributes that make up humanity's propensity to sustain impacts. Much work needs to be done in unravelling and decoding its message.

Given the constant tension that exists between risk amplification and risk mitigation, the balance of these two tendencies determines aggregate vulnerability and the scope of future losses. The resolution of these factors for individual hazards and particular places poses a considerable challenge, but one that holds the key to a deeper understanding of the phenomenon of natural disaster. Disaster studies involve a distinctive amalgam of academic and practical considerations, theoretical and applied concepts, social and physical sciences, natural and technological phenomena, and structural and non-structural mitigation methods. The field has benefited from the tension of opposites created by these dualities, but its development has been held back by the contradictions that they imply.

How, then, can we view natural catastrophes, which are formed out of the tension of opposites, in a unified manner? One way is to consider them in terms of their *context*, as this offers a basis for analyzing disaster in relation to current developments and preoccupations, and with respect to their variation from place to place (Mitchell et al. 1989). The term can be interpreted in various ways: the economic context of international capital, the strategic context of armed conflict, the cultural context of ethnic self-identification, and the technological context of mass communication. One might add the socio-economic context of

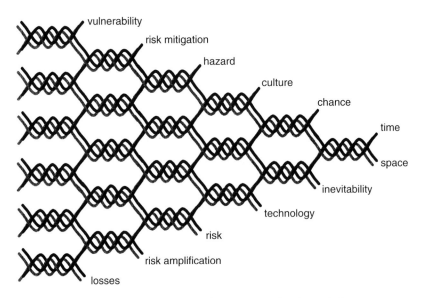

Figure 9.8 The "DNA" of disasters: a series of opposing tendencies and phenomena.

vulnerability and coping mechanisms, the political context of motivation to implement disaster prevention, mitigation and preparedness measures, and the environmental context of impacts and adjustments to them – all of these influence the pattern of disasters and help account for changes. All have undergone radical transformation over the past two decades. The challenge for the future of natural-disaster studies is to be able adequately to draw the connections between different contextual aspects of events as they occur, and to compare events that take place in different contexts.

Passing from the theoretical to the practical, at the time of writing (early 1999) an evaluation of the International Decade for Natural Disaster Reduction would be premature, but certain interim observations can be made. One of the stated aims of the Decade is to reduce disaster impacts by half (US National Research Council 1987; Smith 1996: 342–6; Tobin & Montz 1997: 342–8). In this, success has been patchy, but overall losses continue to rise. In other respects, the Decade set out to abate the toll of disasters by stimulating research on them and fostering its application throughout the world. The primary role of the United Nations, its sponsor, has been as organizer and catalyst for both national endeavours and international collaboration. Many nations have risen to the challenge and have formulated programs for emergency preparedness and catastrophe mitigation. International collaborative efforts have blossomed in various fields associated with disaster reduction. In this respect the Decade has already succeeded, as levels of training, preparation and mitigation have risen steadily since it began. But, on the other hand, vulnerability and losses have continued to increase, despite the best efforts of the mitigators. A pessimist would argue that the magnitude of the problem is so great that the Decade has made only a small dent in it. One could also argue that persistent trends in demography, urbanization and the use of hazard zones tend to counteract any gains in protection against disasters. In the end, the Decade will probably be judged by whether it has succeeded in generating enough momentum to outlast it, or in other words whether the institutions and innovations that it is promoting are maintained over future decades.

Finally, let us engage in a brief speculation on future developments. The overall impact of disasters in the next century may be dominated either by relatively low-consequence high-frequency events, such as recurrent droughts and floods, or by high-consequence infrequent events, such as major earthquakes or massive volcanic eruptions. Past history gives little clue about which of these will be more salient, as the human conditions for disaster have changed considerably, and as very high-magnitude geological events have recurrence intervals that fit in with human timescales and thus render them easily predictable. In the industrialized countries and parts of the developing world, the information technology revolution will probably stimulate a wide variety of creative responses to the monitoring, forecasting and management of disasters. It remains to be seen how far such advances will be extended to the world's poorer countries, which currently lack the infrastructure for implementation. Throughout the world, but especially in the least-developed nations, the pressure of population and the scourge of environmental degradation will almost certainly exacerbate the problem of natural disasters. Momentous changes may also be wrought by processes of global change (Bruce 1993, 1994), which could have extraordinary human consequences in terms of, for example, increased flooding and more powerful hurricanes. One only hopes that adversity will stimulate human ingenuity, as it has in the past, to devise new solutions to the disasters problem.

Conclusion – a model

The times we live in are characterized by pervasive uncertainty and a constantly accelerating pace of change. We have become perilously accustomed to superficial evaluation and rapid reaction, to the jargon of technology and the falsehoods of advertising. Under these circumstances it is difficult to live up to the old adage *finis coronat opus*,[3] for the current milieu does not offer a very firm basis for enduring conclusions. Yet neither should one suffer a failure of nerve. Hence, to conclude, I will attempt to summarize the question of disasters and contemporary social change in terms of a model.

We interpret events and history in the light of our current preoccupations. These can alter significantly with the passage of time and the process of gradual change in human relationships. The latter are heavily influenced by shifting patterns of dominance and submission: for instance, colonialism, postcolonialism and neocolonialism have all had an impact on race relations, and on social, economic and military stability, especially in the tropics. In its many forms, technology has begun to free women from subservience, which has led to a prolonged debate over gender roles and a process of more or less profound alteration in relationships between the sexes, a phenomenon that has spread in varying degrees right around the world. At the global scale and for the time being, capital has triumphed over labour, and capitalism over state-sponsored socialism. This has been accompanied by a series of ideological justifications about the value of free-market trading, whereas those who would question its ethics have been quietened by raising the spectre of mass unemployment. Given the accelerating speed of global economic integration, few alternatives exist;

3. Latin: the end crowns the work.

at the same time, technological change has forced the pace by spreading re-dundancies in traditional industries, or by permitting the relocation of labour-intensive manufacturing to areas where workers can be exploited more suc-cessfully. Harsh realities have focused attention on the imperatives of economic survival in a world whose old certainties have been swept away by the relentless tide of change. Parallel economies have evolved, such as the black market and the drugs trade. Success has become more overtly associated with corruption. Democracy has come to mean manipulation, which is either more or less subtle according to the means adopted.

Increasingly, economic, social and moral justice are administered – or at least heavily influenced – by plebiscite, thanks to the power of televised images. As the English theologian Alcuin (*c.* 735–804) wrote in his Epistles, *Nec audiendi sunt qui solent docere, "Vox populi, vox dei"; cum tumultuositas vulgi semper insaniæ proxima est* ("Nor should we listen to those who say 'The voice of the people is the voice of God,' for the turbulence of the mob is always close to insanity"). Yet 1200 years later, populism has been shown to generate immense rewards for those who can master it, although decidedly not by means of endowing the com-mon man with wisdom.

Herein lies a paradox. Honest leaders are torn between the need to involve the people in decisions and the problems of facing them with more complex issues than they can be expected to understand. Less honest ones have no such scruples; the information age is a distinctly mixed blessing, in that it offers two faces: the quest for truth (a phenomenon that is so often awkward, indetermi-nate, inconclusive, or excessively complex) and the diffusion of what is euphe-mistically known as "disinformation" and more accurately as propaganda or mendacity. Any attempt to cross-check news reports will reveal that the uglier face is surprisingly common (Goltz 1984, Adams 1986, Anderson 1997).

All in all, we now live in what is perhaps the least existential of all ages. It is the age of abstraction. In the West, at least, there has been a blurring of the distinction between recreation and reality, between art and entertainment, between fact, theory and fantasy. It is a dangerous situation, and one that provides fertile opportunities for cynical manipulation. Idealism seems to have gone out of fashion or, more precisely, it is lauded when it can be interpreted in monetary terms as a commodity, when it can be bought and sold. There has been a general failure to recognize that the immense potential of technology as a solver of problems does not extend to fixing human relationships. Happiness is not a video game. In the end, escapism provides no way out.

Paradoxically, the world is fast learning more about itself and sharing that knowledge more readily, but it is becoming more polarized rather than more united. The human race is increasing its degree of interaction but is becoming more fragmentary.[4]

These are the constraints under which we collectively interpret history – history from the past, or history in the making. It is a moral, theological, philosophical, social and ecological muddle. Yet both history and culture are essential keys to the understanding of disaster. Social and technological change are rendering them progressively more difficult to interpret. The rise of the tele-vised image, at once both pervasive and invasive, gives us a *deceptio visus*[5] in which it is easy to think we have grasped reality but very hard to demonstrate it. Curiously, it may be the least existential age, but it is also dangerously close to becoming the age of illusion, in which misery and achievement, the agony and the ecstasy of the human condition, are all too easily compressed into the 60 cm of the cathode ray tube and so lose their full meaning: history, events, as spectacle; trauma as drama. In modern communication systems, immediacy is a relative term. We have learned to accept technology, but not to assimilate it properly, and this has profound implications for how we interpret disaster.

Technological change has divided human societies into communal and indi-vidual kinds, although most fall along a continuum between these two extremes and therefore have elements of both, usually differentiated by social stratum. Communal societies are traditional and are characteristic of low levels of acqui-sition and use of technology. People interact casually and informally rather more than they do in the individualistic kind. The latter is best exemplified by the suburban home, with its electronic entertainments and cars in the garage. In extreme cases it is not even necessary to know one's neighbours: here, people choose their levels and locations of interaction, and sometimes they choose not to interact at all. As the collective reaction to disaster has traditionally been to socialize it, considerable changes can be expected in the way that communities face up to future impacts.

The democracy that is so energetically promoted as a modern ideal would not have been recognizable to the ancient Greeks. It subsists upon huge injec-tions of capital, constant stage management of its protagonists, and generally low levels of public involvement and participation. Many people vote on the basis of whim, fashion or prejudice, rather than firm principle. They vote for what is attractive rather than what is utilitarian. For this reason, natural disaster mitigation is seldom a political card that is worth playing, for it depends on prudence and wisdom, not show; it usually produces unremarkable results, not quick transformations.

Disasters are a window on this contorted world of ours, for they expose the inner workings of society to scrutiny. What we find is a situation in which old

---

4. This is not entirely a bad thing, as the previous unity was mostly imposed by colonial and
   ideological domination.
5. Latin: optical illusion.

sources of vulnerability are being actively mitigated and new ones are springing up. On the positive side, the advancement of learning and spread of information have helped foster new cultures of self-protection against natural catastrophe. There is now enough knowledge of the events and mitigation measures to protect ourselves adequately, which we can do if the resources, socio-economic organization and political will are there. But on the negative side, new pressures are helping to perpetuate the old sources of vulnerability, regardless of the will to ameliorate them. Social, political, economic and military instability are the symptoms; exploitation and lust for power or wealth are the causes. In almost all cases, the antecedents of the problem are clearly evident in history, in past subjugations, ethnic tensions, nationalistic fervours, or economic conquests. The outcomes are found in the decline of traditional coping mechanisms, the rising toll of casualties and losses, and the cumulative weakening of society by recurrent disaster, all of which are set in a matrix of injustice and inequality, the bane of the modern age. At the same time, new sources of vulnerability are appearing. Some of these are *overt*, such as the rise of fixed capital in risk zones, and others are *intrinsic*, such as increasing dependence on potentially fallible technologies.

Given the momentous transformations in human societies and their environments that have occurred over the course of the twentieth century, one cannot understand modern catastrophes fully by using theories that were developed years ago under different social and cultural conditions (e.g. Burton et al. 1978). Even where such conditions are not explicit in the theories, they will have exerted a subtle and pervasive influence on them. One must therefore make theory adaptable to prevailing conditions by examining the context of disaster and its relationship to that milieu. This requires that a suitable definition of context be made, a work that is fraught with difficulty as the phenomena that it must characterize are complex and multifarious. As a result, it is probably too early to talk in terms of overall theories, which in any case require patient and sustained investigation of their validity. Let us therefore venture some tentative hypotheses as a first approximation in order to pick out a few of the characteristics that seem most likely to influence the phenomenology of disasters.

In many modern disasters, context merges with consequence. For example, many students of this field have argued that vulnerability (consisting of marginalization, poverty, malnutrition, and so on), rather than any geophysical agent, is the real cause of disaster (Blaikie et al. 1994). Looked at another way, one may hypothesize the following. The increasing competitiveness of the modern world tends towards a situation in which there are winners and losers, rather than true collaborators in development and mitigation. Increased interaction leads to fragmentation, rather than unification, increased information flow to ideological polarization. Hence, the increased degree of personal choice brought about by economic free-for-all and technological advances leads to moral vacuity.

In such a situation, disasters lose the special significance given to them by the uniqueness of the suffering and losses that they cause. A world increasingly obsessed by materialism is undergoing depreciation in the intrinsic value of disasters as opportunities for profound reflection on the direction that humanity is taking. We will probably not have another *Candide* if Lisbon is rocked again by a major earthquake, nor will there be an end to philosophical optimism, for this has largely abated in the *Realpolitik* of modern atom-bomb culture.

In the present-day world, technological hegemony is one of the main instruments of global power. It is used in both a peaceful way, through the spread of material culture, and belligerently, through the sale and use of armaments.[6] A new ideological wave has swept across the human race. Global realignment has spread power struggles where reconciliation might have been hoped for, through the media of nationalism, ultranationalism, liberation fighting and repression. "Ethnic cleansing" and total warfare have re-emerged from history's darker crevices, to the complete detriment of non-combatants. Atrocities have been used both as justification for ideologies and as actions justified by ideology. Liberalism (as manifest in tolerance, but also in laissez faire and permissiveness) has clashed with fanaticism and traditionalism. The growth of towns and cities has created urban proletariats on a scale unprecedented in human history. Unemployment and free trade have once again become weapons in the ceaseless struggle of capital to divide and rule over labour. Drugs have transformed social mores through injections of dangerous hedonism, and economic mores with illegal and repressive shadow economies. Information overload and obsession with technology have caused a moral emptiness and lack of direction, counterbalanced by the strengthening of more rigid and repressive traditional moralities.

At the same time, education and training, immunization and healthcare have made great strides. On one level, more basic necessities are available than ever before; on the other there is less and less excuse for ignorance about how the world functions outside one's immediate realm of experience. Many of us have become so accustomed to this that we tend to forget what low levels of international understanding prevailed at a popular level even as late as the 1950s. The opportunity to see the world as the home of humanity as a whole, rather than as composed of one's own nation and a host of incomprehensible foreigners, was until very recently the privilege of a tiny elite, not the prerogative of all educated citizens.

For long it has been tempting to see disasters as an issue that is separate from all this, as interruptions to societal processes, rather than as social processes themselves. In reality, they can be understood only by comprehending the

---

6. It is remarkable how resilient arms sales are: despite momentous fluctuations in the global economy, in regional politics, and in ideologies, they continue unabated.

unique mix of universal and particular factors that govern the present state of society in the areas that they commonly affect. Modern life tends to obscure the weight of history under a veneer of modernity that points towards cultural uniformity. This is partly an illusion that masks subtle cultural differentiation with deep historical roots. But it does mean that the same popular tastes, fashionable icons and commercial mechanisms can be found almost anywhere. This leads to the paradox of human memory: we have short memories for events – our **ostensible memory** – and need to be reminded of them; but we have long memories for their cumulative effect – our **innate memory** – which is culturally encoded into our responses, perceptions and behaviour patterns.

As has been noted elsewhere, disasters are part of social reality and in this sense are recurrent events that are to be expected even if they cannot be predicted. But they are also harbingers of change. For example, they are often entry points for socio-economic influence, which is achieved via aid, trade or direct penetration of power structures. Again, rather than leading to unexpected changes, natural catastrophes usually cause a worsening of the current predicament; obviously, death tolls are great where risk levels are high, which is where something is wrong. They are not exceptional events; rather, they are repetitive enough to dull the world's capacity to produce an exceptional response.

Like the Greek goddess Tyche (the Roman Fortuna), natural disasters symbolize both the malevolent and the benevolent faces of human destiny: malevolent because death and destruction prevail during the impact, and exploitation and manipulation may appear afterwards; benevolent because of the solidarity of relief efforts and the stimulus provided by efforts to reconstruct what has been destroyed. It is a classic case of the tension of opposites, the same force that powered the creativity of the Baroque period and continues to impart a rich chiaroscuro to the pattern of current events.

Therefore, let us characterize modern life in terms of its opposites (the **Fortuna hypothesis**). I have grouped these into four categories, which are neither mutually exclusive nor wholly exhaustive, but which together constitute a sort of compass of human destiny (Fig. 9.9). Nothing about its ingredients is inherently new, and each has been inherited and developed over the course of human history to the current point. We will now examine them and then discuss their influence on the course of modern disasters.

The Fortuna hypothesis

Ideocentrism – faith or belief in concepts – can be juxtaposed with material culture, although it does not necessarily exclude other forms of materialism, such as the dialectical kind. We all possess some degree of ideocentrism, according

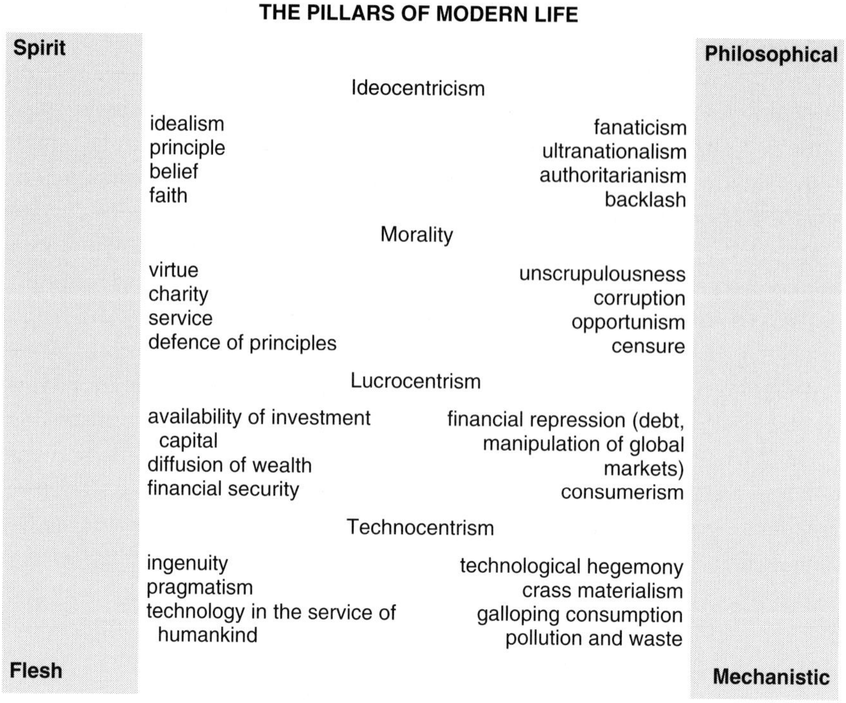

THE PILLARS OF MODERN LIFE

Spirit		Philosophical

Ideocentricism

idealism	fanaticism
principle	ultranationalism
belief	authoritarianism
faith	backlash

Morality

virtue	unscrupulousness
charity	corruption
service	opportunism
defence of principles	censure

Lucrocentrism

availability of investment capital	financial repression (debt, manipulation of global markets)
diffusion of wealth	
financial security	consumerism

Technocentrism

ingenuity	technological hegemony
pragmatism	crass materialism
technology in the service of humankind	galloping consumption
	pollution and waste

Flesh		Mechanistic

Figure 9.9 The pillars of modern life.

to intelligence, education and personal development, but it is manifest in extreme form among people who are intensely religious or are fanatical about some cause. If pushed far enough it can cause an accession of principle or dogma, and when imposed on unwilling subjects it takes the form of authoritarianism.

Morality is the outgrowth of ideocentrism and may take the form of belief in ideals or principles, or faith in codes imposed by a higher power. Yet it may also lead to strains of fanaticism and authoritarianism, including ultranationalism and xenophobia. Morality in this case refers to the guiding principles of public service, but also to their alter ego, the vices of corruption and opportunism. In the present-day world it is much influenced by the mind-set induced by economics and finance.[7] Whereas this confers the benefit of wealth and the security of investment upon society, it also leads to both subtle and overt forms of manipulation, of individuals by consumerism and of governments by the rigours of trade and international debt. We may call this **lucrocentrism**. Lastly, **technocentrism** has greatly increased the horizons of human perception and

---

7. Some 87 per cent of international capital transfers take place for purely speculative reasons, which has led observers to nickname the world of global finance the "lucrosphere".

action, but at the expense of mechanizing values, of insulating people from the dilemmas of moral choice.

The characteristics of these four traits can be linked to the problem of natural disasters as shown by the examples given in Table 9.1. Hence, charity is represented by support of relief appeals and service by voluntary work during emergencies, and unscrupulousness is exemplified by expropriation of relief supplies and corruption by failure to observe building codes and zoning laws that are designed to protect people against hazards. The process of analyzing disasters then becomes one of identifying the most salient opposing tendencies in each of the four categories and assessing the degree to which each one dominates over its opposite.[8]

In synthesis, the context of natural disasters is made up of current conditions and historical factors that are the result of both chance and the preconditioning induced by cultural development. Thus, the influence of the moment (or, if one prefers, the force of circumstance), is superimposed on an historical and cultural organism, whose roots may be deep and heavily ramified. Context is only partly a predeterminant of response to disasters, and hence one should not be too deterministic about it. Moreover, it represents both constraint and opportunity. It imposes limitations on the ability to react to disasters that are both physical (that is, environmental) and social, but it also imparts strength to beleaguered victims, helps clarify situations of dreadful uncertainty, and gives a sense of identity that aids the assumption of personal roles and the promotion of communal solidarity. A slightly circular relationship exists, for disaster adds a small quantity to both history and its own future context.

As society evolves, so it redefines its relationship with formative events, both those of the past and those to come. Hence, I contend that the way to a deeper understanding of natural catastrophe lies in the study of which among the universal factors of human existence is most significant in the context of the moment and under the cumulative duress of history, and which trait counterbalances it.

The approach described in this book has shown abundantly that natural disasters are complex and multifaceted events. Thus, one cannot expect to achieve a complete description of them in a single conceptual model. Many of the fundamental characteristics can be assessed in terms of 16 different scales, which describe aspects of hazard, risk, vulnerability and disaster impact (Table 9.2). The possible combinations of these are so numerous that in these terms each single event is unique, however similar to other events it may seem. Nevertheless, it is possible to put together a single diagram that links the common elements of disaster together in terms of the time periods, effects, contexts and

---

8. This is an outgrowth of the connections depicted in Figure 9.8, the "DNA" of disaster.

Table 9.1 Characteristics of the pillars of modern life as they affect natural disasters and their mitigation

	Positive examples
Ideocentrism	
Ideal	The eradication of all disasters
Principle	Prior mitigation is better than post-disaster repair
Belief	Anti-seismic construction methods can substantially reduce earthquake death tolls
Faith	In human ability to achieve a solution to the problem of disasters
Morality	
Virtue	Untiring application of mitigation measures
Charity	Support of relief appeals
Service	Voluntary work during emergencies
Defence of principles	Ensure that anti-disaster construction codes are respected
Lucrocentrism	
Availability of investment capital	Investment in mitigation and preparedness measures
Diffusion of wealth	Greater preparedness against disasters
Financial security	Reserves for use in combating the effects of future disasters
Technocentrism	
Ingenuity	New hazard monitoring systems
Pragmatism	Intermediate technology for disaster mitigation
Technology in the service of humankind	Information technology for disaster management
	Negative examples
Ideocentrism	
Fanaticism	Attacks on convoys of disaster relief supplies
Ultra-nationalism	A negative component of complex emergencies
Authoritarianism	Inflexibility in disaster relief operations
Backlash	Repression instead of relief
Morality	
Unscrupulousness	Expropriation of relief supplies
Corruption	Failure to observe building codes and zoning laws
Opportunism	Exploitation of disaster situations for personal gain
Censure	Use of disaster situations to pursue moral crusades
Lucrocentrism	
Financial repression	Failure to mitigate as a function of poverty induced by neocolonialism
Consumerism	Disaster as a form of accelerated consumption of goods
Technocentrism	
Technological hegemony	Unequal distribution of the mitigative benefits of technology
Crass materialism	Disaster mitigation takes second place to acquisition of consumer durables
Galloping consumption	High cost of disasters as consumers of resources
Pollution and waste	Dumping of obsolete goods in disaster areas under the guise of supplying relief materials

Table 9.2 Sixteen scales of risk, vulnerability, hazard and disaster impact.

Strong	**Hazard**	*Weak*
	high magnitude ↔ low magnitude	
	high frequency ↔ low frequency	
	well perceived and specified ↔ poorly perceived and specified	
	well mitigated ↔ unmitigated	
	Risk and vulnerability	
	high vulnerability ↔ low vulnerability	
	high exposure ↔ low exposure	
	high risk ↔ low risk	
	imminent risk ↔ latent risk	
	Disaster impact	
	high probability ↔ low probability	
	highly repetitive ↔ single and unique	
	high or concentrated energy ↔ low or diffuse energy	
	direct ↔ indirect or secondary	
	slow impact ↔ sudden impact	
	long duration ↔ short duration (instantaneous)	
	spatially extensive ↔ localized	
	high consequence ↔ low consequence	

influences discussed in this book (Fig. 9.10). This enables one to redefine the causal relationship at the heart of disaster in the following manner:

Extreme geophysical events

result in a

context

that impresses itself cumulatively on

human history and culture

and helps determine the nature of

adaptation to risk and disaster

and influences the toll of

casualties and damage

A short coda will now be offered as a means of presenting a last refinement to the hypothesis.

Well, so what?

In the modern world there is a pervasive tendency to pay lip service to the value of theory, but really to measure things in the light of their current applications. This is often taken to mean whether they make money or save money. It has been stated many times that the value of knowledge cannot be estimated in terms of its current utility, for some of the most obscure discoveries of the past have

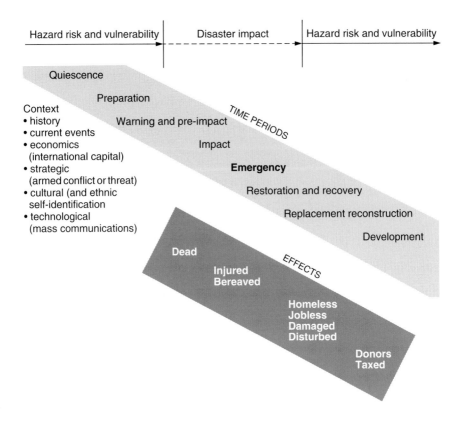

Figure 9.10 Integration of variables that describe disaster processes (see also Table 9.2).

turned out to be fundamental when connected to new problems, accomplishments or technologies. In reality it is not quite so simple, for the process of accumulating useful or edifying knowledge would be extremely inefficient if it were entirely random. Nevertheless, if the development of the sciences and arts is not to be stultified by convention, it must allow for the unexpected.

The health of disaster studies depends on a successful marriage between theory and practice, academic scholarship and field experience. More than in most other disciplines (with the notable exception of the medical ones) pragmatism really does matter, as there is no place for theory that can never be applied.

The main argument of this book is not directed towards solving the everyday problems of emergencies and their management. Yet there is room for theory and abstract study even here. In a simple way, the following case illustrates the point. Local people in an area that recently suffered a damaging earthquake observed scientists unload equipment from a van belonging to a volcanological institute. No-one sought to question these volcanologists and thus the observers were unaware that they had been drafted in to help with a seismological monitoring effort, an exercise in which they were well versed. The story was told to a passing journalist and an article duly appeared in the newspapers entitled "Is a volcanic eruption imminent?" As a result, there was considerable agitation among the earthquake survivors. Yet the area had no history of volcanism and no prospect of its future occurrence. A little learning might have solved the problem and avoided the need to waste time on denials and rebuttals.

Although this book is not a practical manual, I shall nevertheless dedicate a few words to the question of how the ideas it expresses might be utilized in an applied sense. However, I do not propose to carry this process very far, as that ought to be the subject of a different work.

As others have noted (Kelly 1996), emergencies are complex phenomena, regardless of where, when and how they occur. The art and science of managing them are still to some extent hit-and-miss affairs. Improvisation can be reduced to a minimum by training and practice, but it cannot entirely be eliminated. On the other hand, although knowledge does not guarantee power over natural catastrophe, it is a prime requisite of disaster prevention. Modern approaches to emergency actions and mitigation place great emphasis on information management. If only they were to give equal weight to the *quality* of information to be managed. It is here that theory can help, by clarifying the phenomena and processes at work, and by indicating which aspects should be concentrated on in order to solve problems.

An analysis of modern tendencies can help indicate the emerging strengths and weaknesses of society in relation to its ability to tackle the threat and impact of disasters. Once again, we encounter the tension of opposites: technology has furnished us with the means to acquire information, but it has not so readily given us the ability to interpret it. Indeed, technology in the service of commerce has to some extent destroyed traditional coping mechanisms and rendered people pliant and exploitable, rather than independent and self-reliant. As a result, the messages that reach the general public must be tailored to the medium that conveys them and couched in the idioms known to consumers of that medium. If things are to be different, then the whole attitude of society must be altered, not merely its approach to natural disasters.[9]

In synthesis, before planning for disasters, we should learn to see things as they are, not as they were, although this does of course require assiduous study of history and the evolution of human cultures, both of which contain the seeds of modernity. Once this has been accomplished, one can make a further modification to the basic causal process of disasters, as follows:

---

9. Although the task may seem hopeless, there are sometimes fortuitous circumstances, although they may be tragic ones as well. For example, the Chernobyl nuclear accident of 1986 delivered a *coup de grace* to the propagandists of the nuclear industry and forced the public to take a more realistic view of the risks associated with nuclear power (Mould 1992).

Human culture and society
respond to the impact of
extreme geophysical events
and
the forces of socio-economic change
and together these three factors determine the toll of
casualties and damage

To achieve a greater practical understanding of hazard, vulnerability, risk and disaster, we must therefore study the extreme natural events and the evolution of society as joint inputs to the process of catastrophe. They can be compared to one another in terms of the strength and pattern of their influence on disaster. The positive and negative factors that determine resilience and vulnerability can be sought in the tension of opposites, above all in the process of social evolution. In this respect, much work remains to be done.

Bibliography

Abernathy, A. M. & L. Weiner 1995. Evolving federal role for emergency relief. *Forum for Applied Research and Public Policy* **10**(1), 45–9.

Adams, F. D. 1938. *The birth and development of the geological sciences*. Baltimore: Williams & Wilkins (Dover Reprint, 1957 et seq.).

Adams, W. C. 1986. Whose lives count? TV coverage of natural disasters. *Journal of Communication* **36**(2), 113–22.

Aga Khan, S., et al. 1986. *Refugees: the dynamics of displacement*. London: Zed Books.

Agassiz, L. J. R. 1967. *Studies of the glaciers* [translated by A. V. Carozzi]. New York: Hafner. (Originally published as *Système glaciere*. Paris: Masson, 1840.)

Aghababian, R. V. & J. Teuscher 1992. Infectious diseases following major disasters. *Annals of Emergency Medicine* **21**(4), 362–7.

Albala-Bertrand, J. M. 1993. *The political economy of large natural disasters, with special reference to developing countries*. Oxford: Oxford University Press.

Albritton Jr, C. C. 1980. *The abyss of time: changing conceptions of the Earth's antiquity after the sixteenth century*. San Francisco: Freeman, Cooper.

Alesch, D. J. & W. J. Petak 1986. *The politics and economics of earthquake hazard mitigation*. Monograph 43, Natural Hazards Research and Applications Information Centre, University of Colorado, Boulder.

Alexander, D. E. 1980. The Florence floods: what the papers said. *Environmental Management* **4**(1), 27–34.

—— 1981. Disaster in southern Italy, November 1980. *Geographical Magazine* **53**(9), 553–61.

—— 1982a. Leonardo da Vinci and fluvial geomorphology. *American Journal of Science* **282**, 735–55.

—— 1982b. *The earthquake of 23 November 1980 in Campania and Basilicata, southern Italy*. Report, International Disaster Institute, London.

—— 1984. Housing crisis after natural disaster: the aftermath of the November 1980 southern Italian earthquake. *Geoforum* **15**(4), 489–516.

—— 1985. Death and injury in earthquakes. *Disasters* **9**(1), 57–60.

—— 1986a. *Disaster preparedness and the 1984 earthquakes in central Italy*. Working Paper 55, Natural Hazards Research and Applications Information Centre, University of Colorado, Boulder.

—— 1986b. Northern Italian dam failure and mudflow, July 1985. *Disasters* **10**(1), 3–7.

—— 1986c. Dante and the form of the land. *Annals of the Association of American Geographers* **76**(1), 38–49.

—— 1988. Valtellina landslide and flood emergency, northern Italy, 1987. *Disasters* **12**(3), 212–22.

—— 1989a. Urban landslides. *Progress in Physical Geography* **13**(2), 157–91.

—— 1989b. Extraordinary and terrifying metamorphosis: on the seismic causes of slope instability. In *History of geomorphology*, K. J. Tinkler (ed.), 127–50. Boston: Unwin Hyman.

251

——1989c. Spatial aspects of earthquake epidemiology. In *Proceedings of the International Workshop on Earthquake Injury Epidemiology for Mitigation and Response* (Johns Hopkins University, Baltimore, July 1989), P82–P94. Washington DC: National Academy Press.

——1989d. Preserving the identity of small settlements during post-disaster reconstruction in Italy. *Disasters* **13**(3), 228–36.

——1990. Behaviour during earthquakes: a southern Italian example. *International Journal of Mass Emergencies and Disasters* **8**(1), 5–29.

——1991a. Natural disasters: a framework for research and teaching. *Disasters* **15**(3), 209–226.

——1991b. Applied geomorphology and the impact of natural hazards on the built environment. *Natural Hazards* **4**(1), 57–80.

——1991c. Information technology in real time for monitoring and managing natural disasters. *Progress in Physical Geography* **15**(3), 238–60.

——1992. On the causes of landslides: human activities, perception and natural processes. *Environmental Geology and Water Sciences* **20**(3), 165–79.

——1993. *Natural disasters*. London: UCL Press.

——1994a. Il tempo e lo spazio nello studio dei disastri. In *Eventi naturali oggi: la geografia e altre discipline*, G. Botta (ed.), 23–40. Milan: Cisalpino Editrice.

——1994b. The Farakka Barrage and its effects on the geology of the Bengal Basin. *South Asia Forum Quarterly* **7**(2), 1–3.

——1995a. A survey of the field of natural hazards and disaster studies. In *Geographical information systems in assessing natural hazards*, A. Carrara & F. Guzzetti (eds), 1–19. Dordrecht: Kluwer.

——1995b. Changing perspectives on natural hazards in Bangladesh. *Natural Hazards Observer* **19**(6), 1–2.

——1995c. Armenian earthquake leads to calls for building reform. In *Great events from history II: ecology and the environment*, volume 5: *1985–1994*, F. N. Magill (ed.), 1926–31. Pasadena, California: Salem Press.

——1995d. Newspaper reporting of the May 1993 Florence bomb. *International Journal of Mass Emergencies and Disasters* **13**(1), 45–65.

——1995e. Panic during earthquakes and its urban and cultural contexts. *Built Environment* **21**(2/3), 171–82.

——1995f. Fires devastate communities in southern California. In *Great events from history II: ecology and the environment*, volume 5: *1985–1994*, F. N. Magill (ed.), 2108–13. Pasadena, California: Salem Press.

——1996. The health effects of earthquakes in the mid-1990s. *Disasters* **20**(3), 231–47.

——1999. Earthquakes and vulcanism. In *Applied geography: principles and practice*, M. Pacione (ed.), 66–82. London: Routledge.

Alexander, D. E. & L. Coppola 1989. Structural geology and the dissection of alluvial fan sediments by mass movement: an example from the southern Italian Apennines. *Geomorphology* **2**(4), 341–61.

Alexander, Donald 1990. Bioregionalism: science or sensibility. *Environmental Ethics* **12**, 161–73.

Alisio, G. 1979. *Urbanistica napoletana del settecento*. Bari, Italy: Dedalo Libri.

——1984. *Lamont Young: utopia e realtà nell'urbanistica napoletana dell'Ottocento*. Rome: Officina Edizioni.

Allen, T., A. N. M. Mawson, D. Keen, S. Hutchinson 1991. War, famine and flight in Sudan. *Disasters* **15**(2), 126–71.

Almagià, R. 1912. Studi storici di cartografia napoletana. *Archivio Storico delle Province Napoletane* **37**, 564–92.

——1913. Studi storici di cartografia napoletana. *Archivio Storico delle Province Napoletane* **38**, 3–35, 318–48, 409–440, 639–54.

Ambrose, J. & D. Vergun 1985. *Seismic design of buildings*. New York: John Wiley.

Anderson, A. 1997. *Media, culture and the environment*. New Brunswick, New Jersey: Rutgers University Press.

Anderson, J. G. 1978. On the attenuation of modified Mercalli intensity with distance in the United States. *Bulletin of the Seismological Society of America* **68**(4), 1147–80.

Anderson, J. W. 1967. Cultural adaptation to threatened disaster. *Human Organization* **27**, 298–307.

Anderson, M. B. 1991. Which costs more: prevention or recovery? In *Managing natural disasters and the environment*, A. Kreimer & M. Munasinghe (eds), 17–27. Washington DC: World Bank (Environment Department).

—— 1992. Metropolitan areas and disaster vulnerability: a consideration for developing countries, environmental management and urban vulnerability. In Discussion Paper 168, A. Kreimer & M. Munasinghe (eds). Washington DC: World Bank (Environment Department).

Anderson, P. S. 1994. The Emergency Preparedness Information Exchange Project (EPIX). *Stop Disasters* **22**, 20.

Anderson, W. A. 1969. Social structure and the role of the military in natural disaster. *Sociology and Social Research* **53**, 242–52.

Andrews, L. G. & J. A. Souma 1989. Dialysis relief effort for Armenia. *New England Journal of Medicine* **321**, 264–5.

Antonopoulos, J. 1992. The great Minoan eruption of Thera volcano and the ensuing tsunami in the Greek archipelago. *Natural Hazards* **5**(2), 153–68.

Aptekar, L. 1991. *The psychosocial process of adjusting to natural disasters*. Working Paper 70, Natural Hazard Research and Applications Information Center, University of Colorado, Boulder.

Arboleda, R. A. & R. S. Punongbayam 1991. Landslides induced by the 16 July 1990 Luzon, Philippines, earthquake. *Landslide News* **5**, 5–7.

Aristotle 1950. *De generatione et corruptione* (translated by E. S. Forster). Cambridge, Massachusetts: Harvard University Press.

—— 1952. *Meteorologica* (translated by H. D. P. Lee). Cambridge, Massachusetts: Harvard University Press.

Auf der Heide, E. 1989. *Disaster response: principles of preparation and coordination*. St Louis, Missouri: Mosby.

Autier, P., et al. 1990. Drug supply in the aftermath of the 1988 Armenian earthquake. *Lancet* **335**(9 June), 1388–90.

Bailey, K. D. 1989. Taxonomy and disaster: prospects and problems. *International Journal of Mass Emergencies and Disasters* **7**(3), 419–32.

Baker, S., et al. 1974. The injury severity score: a method for describing patients with multiple injuries and evaluating emergency care. *Journal of Trauma* **14**, 187–96.

Baker, V. R. 1988. Cataclysmic processes in geomorphological systems. *Zeitschrift für Geomorphologie* [Supplementband] **67**, 25–32.

Baker, V. R., Z. C. Kochel, P. C. Patton (eds) 1988. *Flood geomorphology*. New York: John Wiley.

Ball, N. 1979. Some notes on defining disasters: suggestions for a disaster continuum. *Disasters* **3**(1), 3–7.

Barbat, A. H. 1996. Damage scenarios simulation for seismic risk assessment in urban zones. *Earthquake Spectra* **12**(3), 371–94.

Barberi, F., M. Rosi, R. Santocroce, M. F. Sheridan 1983. Volcanic hazard zonation: Mount Vesuvius. In *Forecasting volcanic events*, H. Tazieff & J. C. Sabroux (eds), 149–61. Amsterdam: Elsevier.

Barberi, F., G. Macedonia, M. T. Pareschi, R. Santacroce 1990. Mapping the tephra fallout risk: an example from Vesuvius, Italy. *Nature* **344**, 142–4.

Barker, D. & D. Miller 1990. Hurricane Gilbert: anthropomorphising a national disaster. *Area* **22**(2), 107–116.

Barkun, M. 1977. Disaster in history. *Mass Emergencies* **2**, 219–31.

Barrows, H. H. 1923. Geography as human ecology. *Annals of the Association of American Geographers* **13**, 1–14.

Barton, A. M. 1970. *Communities in disaster: a sociological analysis of collective stress situations*. Garden City, New York: Anchor Books.

Bates, F. L. & C. Pelanda 1994. An ecological approach to disasters. In *Disasters, collective behavior, and social organization*, R. R. Dynes & K. J. Tierney (eds), 145–59. Newark, Delaware: University of Delaware Press.

Baumann, D. D. & J. H. Sims 1974. Human response to the hurricane. In *Natural hazards: local, national, global*, G. F. White (ed.), 25–30. New York: Oxford University Press.

Beatley, T. 1989. Towards a moral philosophy of natural disaster mitigation. *International Journal of Mass Emergencies and Disasters* **7**(1), 5–32.

Beatley, T. & P. Berke 1990. Seismic safety through public incentives: the Palo Alto seismic hazard identification program. *Earthquake Spectra* **6**(1), 57–79.

Beijing Normal University 1992. *Atlas of natural disasters in China*. Beijing: Science Press, for Beijing Normal University and People's Insurance Company of China.

Belardo, S., H. L. Pazer, W. A. Wallace, W. D. Danko 1983. Simulation of a crisis management information network: a serendipitous evaluation. *Decision Sciences* **14**, 588–606.

Belmonte, T. 1989. *The broken fountain* (2nd edn). New York: Columbia University Press.

Belt Jr, C. B. 1975. The 1973 flood and man's constriction of the Mississippi River. *Science* **189**, 681–4.

Belward, A. S., P. J. Kennedy, J-M. Gregoire 1994. The limitations and potential of AVHRR GAC data for continental-scale fire studies. *International Journal of Remote Sensing* **15**, 243–54.

Bender, S. O. 1992. Disaster preparedness and sustainable development. In *Proceedings of the Conference on Science and Technology in the Developing World: Liberation or Dependence?* Indiana Center on Global Change and World Peace, Indiana University, Bloomington, 8–9 October 1992.

Bennett, G. 1979. Reports from China: mass campaigns and earthquakes: Hai Cheng, 1975. *China Quarterly* **77**, 94–112.

Benthall, J. 1993. *Disasters, relief and the media*. New York: St Martin's Press.

Bernal, J. D. 1949. *The freedom of necessity*. London: Routledge & Kegan Paul.

Bernard, E. N. 1991. Assessment of Project THRUST: past, present, future. *Natural Hazards* **4**(2–3), 285–92.

Berry, J. 1969. On cross-cultural comparability. *International Journal of Psychology* **4**, 119–28.

Berz, G. 1988. List of major natural disasters, 1960–1987. *Earthquakes and Volcanoes* **20**(6), 226–8.

—— 1991. Natural disasters and insurance and re-insurance. *Earthquakes and Volcanoes* **22**(3), 99–102.

—— 1992. Losses in the range of US$50 billion and 50 000 people killed: Munich Re's list of major natural disasters in 1990. *Natural Hazards* **5**(1), 95–102.

—— 1994. The insurance industry and IDNDR: common interests and tasks. *Natural Hazards* **9**(3), 323–32.

Bianco, B. 1986. Il consolidamento della frana di Fontivegge in Perugia: problemi tecnico-esecutivi. *Associazione Geotecnica Italiana, 16th National Conference on Engineering Geology, Bologna, 14–16 May 1986*: 101–110.

Bianchi, R., A. Coradini, C. Federico, G. Giberti, P. Lanciano, J. P. Pozzi, G. Sartoris, R. Scandone 1987. Modelling of surface deformation in volcanic areas: the 1970–1974 and 1982–1984 crises of Campi Flegrei, Italy. *Journal of Geophysical Research* **92**(13), 14139–14150.

Binaghi Olivari, M. T., et al. 1980. *Le pietre dello scandolo*. Turin: Einaudi.

Birkland, T. A. 1997a. *After disaster: agenda setting, public policy and focusing events*. Washington DC: Georgetown University Press.

—— 1997b. Factors inhibiting a national hurricane policy. *Coastal Management* **25**(4), 387–403.

Blaikie, P. 1985. *The political economy of soil erosion in developing countries*. Harlow, England: Longman.

Blaikie, P., T. Cannon, I. Davis, B. Wisner 1994. *At risk: natural hazards, people's vulnerability and disasters*. London: Routledge.

Black, R. 1994. Environmental change in refugee-affected areas of the Third World: the role of policy and research. *Disasters* **18**(2), 107–116.

Blocker, T. J., E. B. Rochford Jr, D. E. Sherkat 1991. Political responses to natural hazards: social movement participation following a flood disaster. *International Journal of Mass Emergencies and Disasters* **9**(3), 367–82.

Blunt, A. 1975. *Neapolitan baroque and rococó architecture*. London: Zwemmer.

Bollens, S. A., E. J. Kaiser, R. J. Burby 1988. Evaluating the effects of local floodplain management policies on property owner behavior. *Environmental Management* **12**(3), 311–25.

Bolt, B. A. 1993. *Earthquakes* (3rd edn). New York: W. H. Freeman.

Borchardt, G. 1991. Preparation and use of earthquake planning scenarios. *California Geology* **44**(9), 195–203.

Borton, J. 1993. Recent trends in the international relief system. *Disasters* **17**(3), 187–201.

Boyce, J. K. 1990. Birth of a megaproject: political economy of flood control in Bangladesh. *Environmental Management* **14**(4), 419–28.

Bradbury, M. 1998. Normalizing the crisis in Africa. *Disasters* **22**(4), 328–38.

Brammer, H. 1987. Drought in Bangladesh: lessons for planners and administrators. *Disasters* **11**(1), 21–9.

———1990. Floods in Bangladesh, I: geographical background to the 1987 and 1988 floods. *Geographical Journal* **156**(1), 12–22.

Brand, E. W. 1994. Landslide in Hong Kong during the rainfall event of 4–5 November 1993. *Landslide News* **8**, 35–6.

Brantley, S. R. (ed.) 1990. *The eruption of Redoubt Volcano, Alaska, December 14, 1989 to August 31, 1990*. Circular 1061, US Geological Survey, Washington DC.

Brazee, R. J. 1979. Re-evaluation of modified Mercalli intensity scale using distance as determinant. *Bulletin of the Seismological Society of America* **69**(3), 911–24.

Brislin, R. W. 1980. Cross-cultural research methods. In *Environment and culture*, I. Altman, A. Rapoport, J. F. Wohlwill (eds), 47–82. New York: Plenum.

Bristow, C. S. 1987. Brahmaputra River: channel migration and deposition. In *Recent developments in fluvial sedimentology*, F. G. Ethridge, R. M. Flores, M. D. Harvey (eds), 67–74. Tulsa, Oklahoma: Society of Economic Paleontologists and Mineralogists.

Britton, N. R. 1984. Australia's organized response to natural disasters: the constrained organization of two wildfire settings. *Disasters* **8**(3), 214–25.

Britton, N. R. & J. Oliver (eds) 1997. *Financial risk management for natural catastrophes*. Sydney: Aon Group.

Broad, R. & J. Cavanagh 1993. *Plundering Paradise: the struggle for the environment of the Philippines*. Berkeley: University of California Press.

Brown, J. M. & E. A. Campbell 1991. Stress among emergency services personnel: progress and problems. *Journal of the Society of Occupational Medicine* **41**(4), 149–50.

Browning, J. M. 1973. Catastrophic rock slide, Mount Huascaran, north-central Peru, May 31, 1970. *Bulletin of the American Association of Petroleum Geologists* **57**, 1335–41.

Bruce, J. P. 1993. Natural disasters and global change. *Stop Disasters* **15**, 3.

———1994. Natural disaster reduction and global change. *Bulletin of the American Meteorological Society* **75**(10), 1831–5.

Bruhn, D. & P. Li 1989. Research activities at the Jing Long Shan landslide observation station in southwestern Sicuan, China. *Landslide News* **3**, 14–16.

Brunsden, D. & J. B. Thornes 1979. Landscape sensitivity and change. *Institute of British Geographers, Transactions* **4**(4), 463–84.

Bunge, W. W. 1966. *Theoretical geography* [Lund Studies in Geography, Series C, no. 1]. Lund, Sweden: Gleerup.

Bunin, J. 1989. Incorporating ecological concerns into the IDNDR. *Natural Hazards Observer* **14**(2), 4–5.

Burby, R. J., B. A. Cigler, S. P. French, E. J. Kaiser, J. Kartez, D. Roenigk, D. Weist, D. Whittington 1991. *Sharing environmental risks: how to control governments' losses in natural disasters*. Boulder, Colorado: Westview.

Burton, I. & K. Hewitt 1974. Ecological dimensions of environmental hazards. In *Human ecology*, F. Sargent (ed.), 253–83. Amsterdam: Elsevier.

Burton, I. & R. W. Kates 1964a. The perception of natural hazards in resource management. *Natural Resources Journal* **3**(3), 412–41.

—— 1964b. The floodplain and the sea shore: a comparative analysis of hazard zone occupance. *Geographical Review* **54**, 366–85.

Burton, I., R. W. Kates, G. F. White 1968. *The human ecology of extreme geophysical events*. Working Paper 1, Natural Hazards Research and Applications Information Center, University of Colorado, Boulder.

—— 1978. *The environment as hazard*. New York: Oxford University Press.

—— 1981. The future of hazard research: a reply to William I. Torry. *Canadian Geogapher* **25**(3), 286–9.

—— 1993. *The environment as hazard* (2nd edn). New York: Guilford Press.

Burton, I. & R. Pushchak 1984. The status and prospects of risk assessment. *Geoforum* **15**(3), 463–76.

Butler, D. I. 1997. Selected Internet sites on natural hazards and disasters. *International Journal of Mass Emergencies and Disasters* **15**(1), 197–215.

Butler, D. R., S. J. Walsh, D. G. Brown 1991. Three-dimensional displays for natural hazards analysis using classified LANDSAT Thematic Mapper digital data and large-scale digital elevation models. *Geocarto International* **6**(4), 65–9.

Butzer, K. W. 1989. Cultural ecology. In *Geography in America*, G. Gaile & C. Wilmott (eds), 192–208. Columbus, Ohio: Charles Merrill.

—— 1990. The realm of cultural–human ecology: adaptation and change in historical perspective. In *The Earth as transformed by human action: global and regional changes in the biosphere over the past 300 years*, B. L. Turner II, W. C. Clark, R. W. Kates, J. F. Richards, J. L. Mathews, W. B. Meyer (eds), 685–701. New York: Cambridge University Press.

Cambefort, H. 1978. Foundation performance of Tower of Pisa: discussion. *Proceedings of the American Society of Civil Engineers, Journal of the Geotechnical Engineering Division* **104**, 156–60.

Cannon, T. 1994. Vulnerability analysis and the explanation of "natural" disasters. In *Disasters, development and environment*, A. Varley (ed.), 13–30. Chichester, England: John Wiley.

Carey, S. & H. Sigurdsson 1987. Temporal variations in column height and magma discharge rate during the AD 79 eruption of Vesuvius. *Geological Society of America, Bulletin* **99**, 303–314.

Carozzi, M. 1983. Voltaire's attitude to geology. *Archives des Sciences, Genève* **36**, 1–145.

Carroll, J. M. (ed.) 1983. *Computer simulation in emergency planning*. Simulation Series 11, Society for Computer Simulation, La Jolla, California.

Carta, S. R. Figari, A. Sartoris, E. Sassi, R. Scandone 1981. A statistical model for Vesuvius and its volcanological significance. *Bulletin Volcanologique* **44**, 117–51.

Carter, L. J. 1977. Auburn dam: earthquake hazards imperil $1 billion project. *Science* **197**, 643–7.

Cate, F. H. 1995. *International disaster communications: harnessing the power of communications to avert disasters and save lives*. Washington DC: Annenberg Washington Program.

Changnon, S. A. (ed.) 1996. *The Great Flood of 1993: causes, impacts, and responses*. Boulder, Colorado: Westview.

Chen, Y., J. Liu, L. Chen, Q. Chen, L. S. Chan 1998. Global seismic hazard assessment based on area source model and seismicity data. *Natural Hazards* **17**(3), 251–67.

Cheng Yong, Kam-ling Tsoi, Chen Feibi, Gao Zhenhuan, Zou Qijia, Chen Zhangli (eds) 1988. *The great Tangshan earthquake of 1976: an anatomy of disaster*. Oxford: Pergamon.

Chen Zhiming 1996. Sequential analysis of China's hazards in geoscience. *GeoJournal* **38**(3), 259–63.

Chilingarian G. V., E. C. Donaldson, T. F. Yen (eds) 1995. *Subsidence due to fluid withdrawal*. Amsterdam: Elsevier.

Chomsky, N. 1989. *Necessary illusions: thought control in democratic societies*. Boston, Massachusetts: South End Press.

—— 1994. *Secrets, lies and democracy*. Tucson, Arizona: Odonian Press.

Chorley, R. J., A. J. Dunn, R. P. Beckinsale 1964. *The history of the study of landforms*, vol. I:

geomorphology before Davis. London: Methuen.

Chorley, R. J. & B. A. Kennedy 1971. *Physical geography: a systems approach*. Hemel Hempstead, England: Prentice-Hall.

Choudhury, G. S. & N. P. Jones 1996. Development and application of data-collection forms for post-earthquake surveys of structural damage and human casualties. *Natural Hazards* **13**(1), 17–38.

Clapham, C. 1991. The structure of regional conflict in northern Ethiopia. *Disasters* **15**(3), 244–53.

Clark, K. 1953. *Landscape into art*. London: John Murray.

Clarke, J. N. & A. K. Gerlak 1998. Environmental racism in southern Arizona? The reality beneath the rhetoric. *Environmental Management* **22**, 857–68.

Clout, H., M. Blacksell, R. King, D. Pinder 1994. *Western Europe: geographical perspectives* (3rd edn). Harlow, England: Longman.

Coburn, A. & R. Spence 1992. *Earthquake protection*. New York: John Wiley.

Coleman, J. M. 1969. Brahmaputra River, channel processes and sedimentation. *Sedimentary Geology* **3**, 129–239.

Committee on Earthquake Engineering 1985. *Liquefaction of soils during earthquakes*. Washington DC: National Research Council, National Academy Press.

Committee on Safety Criteria for Dams 1985. *Safety of dams: flood and earthquake criteria*. Washington DC: National Academy of Sciences.

Congalton, R. G. 1991. A review of methods of assessing the accuracy of classification of remotely sensed data. *Remote Sensing of the Environment* **37**, 35–46.

Cooke, R. U. 1984. *Geomorphological hazards in Los Angeles*. London: Allen & Unwin.

Cooke, R. U. & R. W. Reeves 1975. *Arroyos and environmental change in the American Southwest*. Oxford: Oxford University Press.

COPAT 1981. *Bombs for breakfast*. London: Committee on Poverty and the Arms Trade.

Cortner, H. J., P. D. Gardner, J. G. Taylor 1990. Fire hazard at the urban–wildland interface: what the public expects. *Environmental Management* **14**(1), 57–62.

Costa, J. E. 1978. The dilemma of flood control in the United States. *Environmental Management* **2**(4), 313–22.

Costa, J. E. & R. L. Schuster 1988. The formation and failure of natural dams. *Bulletin of the Geological Society of America* **100**, 1054–1068.

Covello, V. T. & J. Mumpower 1985. Risk analysis and risk management: an historical perspective. *Risk Analysis* **5**(2), 103–120.

Crouse, J. & D. Trusheim 1988. *The case against the SAT*. Chicago: University of Chicago Press.

Cuny, F. C. 1983. *Disasters and development*. New York: Oxford University Press.

Cutter, S. L. (ed.) 1994. *Environmental risks and hazards*. Englewood Cliffs, New Jersey: Prentice-Hall.

Dacy, D. C. & H. Kunreuther 1969. *The economics of natural disaster: implications for federal policy*. New York: The Free Press.

Daines, G. E. 1991. Planning, training and exercising. In *Emergency management: principles and practice for local government*, T. E. Drabek & G. J. Hoetmer (eds), 161–200. Washington DC: International City Management Association.

Dammers, C. 1992. Iraq: a disaster for the 1990s? *Disasters* **15**(4), 355–62.

Davies, G. L. H. 1989. On the nature of geo-history, with reflections on the historiography of geomorphology. In *History of geomorphology*, K. J. Tinkler (ed.), 1–10. Boston: Unwin Hyman.

Davis, I. 1978. *Shelter after disaster*. Oxford: Oxford Polytechnic Press.

Davis, M. & S. T. Seitz 1982. Disasters and governments. *Journal of Conflict Resolution* **26**, 547–68.

Davis, N. Y. 1986. Earthquake, tsunami, resettlement and survival in two Pacific Alaskan native villages. In *Natural disasters and cultural responses*. V. H. Sutlive, N. Altshuler, M. D. Zamora, V. Kerns (eds), 123–54. Studies in Third World Societies 36, Department of Anthropology, College of William and Mary, Williamsburg, Virginia.

257

De Bruycker, M., D. Greco, I. Annino, M. A. Stazi, N. De Ruggiero, M. Triassi, Y. P. De Kettenis, M. F. Lechat 1983. The 1980 earthquake in southern Italy: rescue of trapped victims and mortality. *Bulletin of the World Health Organization* **61**(6), 1021–1025.

De Cunzo, M. 1979. *Le ville vesuviane: civiltà del '700 a Napoli, 1734–1799*, 86–105. Florence: Centro Di.

De Filippis, F. 1971. *Le antiche residenze a Napoli*. Cava de' Tirreni, Salerno: Di Mauro.

De Luca, G. 1974. *Napoli, una vicenda: raccolta di scritti* (edited by M. Rosi). Naples: Guida Editori.

Denis, H. 1997. Technology, structure and culture in disaster management: coping with uncertainty. *International Journal of Mass Emergencies and Disasters* **15**(2), 293–308.

Denatale, G. & F. Pingue 1993. Ground deformations in collapsed caldera structures. *Journal of Volcanology* **57**, 19–38.

Denis, H. 1997. Technology, structure and culture in disaster management: coping with uncertainty. *International Journal of Mass Emergencies and Disasters* **15**(2), 293–308.

De Seta, C. 1969. *Cartografia della città di Napoli*. Naples: Edizioni Scientifiche Italiane.

De Seta, C., L. Di Mauro, M. Perone 1980. *Ville vesuviane*. Milano: Rusconi.

Desmarest, Nicolas 1771. Mémoires sur l'origine et la nature du basalte à grandes colonnes polygones, déterminées par l'histoire naturelle de cette pierre, observée en Auvergne. *Mémoires de l'Académie Royale des Sciences, Paris*, 599–670.

De Ville De Goyet, C. 1993. Post-disaster relief: the supply-management challenge. *Disasters* **17**(2), 169–76.

De Ville De Goyet, C., E. Del Cid, A. Romero, E. Jeannee, M. F. Lechat 1976. Earthquake in Guatemala: epidemiological evaluation of the relief effort. *Bulletin of the Pan American Health Organization* **10**, 95–109.

De Ville De Goyet, C. & M. F. Lechat 1976. Health aspects in natural disasters. *Tropical Doctor* **6**, 152–7.

Diaz, H. F. & R. S. Pulwarty (eds) 1997. *Hurricanes: climate and socioeconomic impacts*. New York: Springer.

Di Leva, G. 1996. *Firenze: cronaca del diluvio, 4 novembre 1966*. Florence: Le Lettere.

Dirks, R. 1979. Relief induced agonism. *Disasters* **3**(2), 195–8.

Disaster Prevention and Management: an International Journal 1991–.

Di Stefano, R., et al. 1967. Dissesti nella città di Napoli e loro cause. *Atti dell'VIII Convegno di Geotecnica, Cagliari, 1967*, 217–26. Naples: Edizioni Scientifiche Italiane.

Dobran, F., A. Nerl, M. Tedesco 1994. Assessing the pyroclastic flow hazard at Vesuvius. *Nature* **367**, 551–4.

Dolan, R., B. Hayden, K. Bosserman, L. Isle 1987. Frequency and magnitude data on coastal storms. *Journal of Coastal Research* **3**(2), 245–7.

Doornkamp, J. C. & M. Han 1985. Morphotectonic research in China and its application to earthquake prediction. *Progress in Physical Geography* **7**, 353–81.

Douglas, M. & A. Wildavsky 1982. *Risk and culture: an essay on the selection of technical and environmental dangers*. Berkeley: University of California Press.

Douty, C. M. 1977. *The economics of localized disasters*. New York: Arno Press.

Downs, R. E., D. O. Kerner, S. P. Reyner (eds) 1991. *The political economy of African famine*. Philadelphia: Gordon & Breach.

Drabek, T. E. 1986. *Human system response to disaster: an inventory of sociological findings*. New York: Springer.

——— 1987. *The professional emergency manager: structures and strategies for success*. Monograph 44, Natural Hazards Research and Applications Information Centre, University of Colorado, Boulder.

——— 1989. Disasters as nonroutine social problems. *International Journal of Mass Emergencies and Disasters* **7**(3), 253–64.

——— 1990. *Emergency management: strategies for maintaining organizational integrity*. New York: Springer.

——— 1991. The evolution of emergency management. In *Emergency management: principles and practice for local government*, T. E. Drabek & G. J. Hoetmer (eds), 3–29. Washington DC: International City Management Association.

Drabek, T. E. & G. J. Hoetmer (eds) 1991. *Emergency management: principles and practice for local government*. Washington DC: International City Management Association.

Duckworth, D. H. 1986. Psychological problems arising from disaster work. *Stress Medicine* **2**, 315–23.

Dudasik, S. 1982. Unanticipated repercussions of international disaster relief. *Disasters* **6**(1), 31–7.

Dudley, E. 1988. Disaster mitigation: strong houses or strong institutions? *Disasters* **12**(2), 111–21.

Duffield, J. W. 1980. Auburn Dam, a case study of water policy and economics. *Water Resources Bulletin* **16**, 226–34.

Duffield, M. 1990. From emergency to social security in Sudan, part I: the problem. *Disasters* **14**(1), 187–203.

—— 1996. The symphony of the damned: racial discourse, complex political emergencies and humanitarian aid. *Disasters* **20**(3), 173–93.

Durkin, M. E. 1995. Fatalities, nonfatal injuries, and medical aspects of the Northridge earthquake. In *The Northridge, California, earthquake of 17 January 1994*, M. C. Woods & W. R. Seiple (eds), 247–54. Special Publication 116, Division of Mines and Geology, California Department of Conservation, Sacramento.

Durkin, M. E., C. C. Thiel Jr, J. E. Schneider 1994. Casualties and emergency medical response. In *The Loma Prieta, California, earthquake of October 17, 1989: loss estimation procedures*, S. K. Tubbesing (ed.), 9–38. Professional Paper 1553A, US Geological Survey, Washington DC.

Dury, G. H. 1980. Neocatastrophism? A further look. *Progress in Physical Geography* **4**(3), 391–413.

Dynes, R. R. 1970. *Organized behaviour in disaster*. Lexington, Massachusetts: D. C. Heath.

Dynes, R. R. & K. J. Tierney (eds) 1994. *Disasters, collective behavior and social organization*. Cranbury, New Jersey: Associated University Presses.

Earthquake Spectra 1991. Philippines earthquake reconnaissance report. *Earthquake Spectra* **7**, Supplement A.

Ebert, C. H. V. 1981. The seismic geography of Managua, Nicaragua. *Ecumene* **13**, 59–70.

Ecologist, The 1992. The Earth Summit debacle. *The Ecologist* **22**(4), 122.

Economist, The 1996. Launch of Ariane 5: an elephant explodes. *The Economist* **339**(7969), 87.

Ehlers, M., G. Edwards, Y. Bedard 1989. Integration of remote sensing with geographic information systems: a necessary evolution. *Photogrammetric Engineering and Remote Sensing* **55**(11), 1619–28.

Eidinger, J. 1996a. An overview of the "HAZUS" earthquake loss estimation methodology. In *Second National Workshop on Modeling Earthquake Casualties for Planning and Response: summary of proceedings*, R. A. Olson & D. E. Alexander (eds), 6. Sacramento, California: Robert Olson Associates.

—— 1996b. Palo Alto loss estimate case study. In *Second National Workshop on Modeling Earthquake Casualties for Planning and Response: summary of proceedings*, R. A. Olson & D. E. Alexander (eds), 9–11. Sacramento, California: Robert Olson Associates.

El-Baz, F. 1999. War, environmental impacts. In *The encyclopedia of environmental science*, D. E. Alexander & R. W. Fairbridge (eds), 668–9. Dordrecht: Kluwer.

Ellingwood, B. R., R. B. Corotis, J. Boland, N. P. Jones 1993. Assessing the cost of dam failure. *Journal of Water Research Planning and Management (ASCE)* **119**(1), 64–82.

El-Sabh, M. I. & T. S. Murty (eds) 1988. *Natural and man-made hazards*. Dordrecht: Kluwer.

Emmi, P. C. & C. A. Horton 1993. GIS-based assessment of earthquake property damage and casualty risk, Salt Lake County, Utah. *Earthquake Spectra* **9**(1), 11–33.

—— 1995. A Monte Carlo simulation of error propagation in a GIS-based assessment of seismic risk. *International Journal of Geographical Information Systems* **9**(4), 447–61.

Ente per le Ville Vesuviane 1981. *Le ville vesuviane.* Naples: Ente per le Ville Vesuviane.

Erikson, K. T. 1976. *Everything in its path: destruction of community in the Buffalo Creek flood.* New York: Simon & Schuster.

Esteva, L. 1988. The Mexico City earthquake of September 19, 1985: consequences, lessons and impact on research and practice. *Earthquake Spectra* **4**(3), 413–26.

Fahmi, K. J. & J. N. Alabbasi 1989. Seismic intensity zoning and earthquake risk mapping in Iraq. *Natural Hazards* **1**(4), 331–40.

Fan Yuchen 1992. Disaster relief in China. *Disasters* **15**(4), 379–81.

Faupal, C. E. 1985. *The ecology of disaster: an application of a conceptual model.* New York: Irvington.

Fearnside, P. M. 1989. Brazil's Balbina Dam: environment versus the legacy of the Pharaohs in Amazonia. *Environmental Management* **13**(3), 401–423.

FEMA 1988. *Rapid screening of buildings for potential seismic hazards: a handbook.* Washington DC: Federal Emergency Management Agency.

Fielden, B. 1980. Earthquakes and historic buildings. *Proceedings of the Seventh World Conference on Earthquake Engineering* **9**, 213–26.

Fischer, H. W. III 1998. The role of the new information technologies in emergency mitigation, planning, response and recovery. *Disaster Prevention and Management* **7**(1), 28–37.

Form, W. H. & C. P. Loomis 1956. The persistence and emergence of social and cultural systems in disasters. *American Sociological Review* **21**, 180–85.

Formica, C. 1966. Il Vesuvio: studio antropogeografico. *L'Universo* **46**, 525–8.

Foster, H. D. 1976. Assessing disaster magnitude: a social science approach. *Professional Geographer* **28**(3), 241–7.

——— 1980. *Disaster planning: the preservation of life and property.* New York: Springer.

França, J-A. 1983. *Lisboa pombalina e o illuminismo.* Lisbon: Bertrand.

Freeberne, M. 1962. Natural calamities in China, 1949–61: an examination of the reports originating from the mainland. *Pacific Viewpoint* **3**, 33–72.

Frerks, G., T. J. Kliest, S. J. Kirkby, N. D. Emmel, P. O'Keefe, I. Convery 1995. A disaster continuum? *Disasters* **19**(4), 362–6.

Freudenburg, W. R. 1988. Perceived risk, real risk: social science and the art of probabilistic risk assessment. *Science* **242**, 44–9.

Fritz, C. E. 1957. *Convergence behavior in disasters: a problem in social control.* Special Report, National Academy of Sciences and National Research Council, Washington DC.

Fritz, C. E. & E. S. Marks 1954. The NORC studies of human behaviour in disaster. *Journal of Social Issues* **10**(3), 26–41.

Geipel, R. 1990. *The long-term consequences of disasters: the reconstruction of Friuli, Italy, in its international context, 1976–1988.* Heidelberg: Springer.

Gheradi, S. 1998. A cultural approach to disasters. *Journal of Contingencies and Crisis Management* **6**(2), 80–83.

Gillespie, C. C. 1951. *Genesis and geology.* Cambridge, Massachusetts: Harvard University Press.

Gillespie, D. F. 1988. Barton's theory of collective stress is a classic and worth testing. *International Journal of Mass Emergencies and Disasters* **6**(3), 345–62.

Giustiniani, L. 1969 [reprint]. *Dizionario geografico ragionato del Regno di Napoli (1797–1805).* Bologna: Forni.

Glacken, C. J. 1967. *Traces on the Rhodian shore: nature and culture in Western thought from ancient times to the end of the eighteenth century.* Berkeley: University of California Press.

Glantz, M. H. 1982. Consequences and responsibilities in drought forecasting: the case of Yakima, 1977. *Water Resources Research* **18**, 3–13.

Gleijeses, V. 1980. *Ville e palazzi vesuviani.* Naples: Società Editrice Napoletana.

Godschalk, D. R. 1991. Disaster mitigation and hazard management. In *Emergency management: principles and practice for local government,* T. E. Drabek & G. J. Hoetmer (eds), 131–60. Washington DC: International City Management Association.

Goltz, J. D. 1984. Are the news media responsible for the disaster myths? A content analysis of emergency response imagery. *International Journal of Mass Emergencies and Disasters* **2**(3), 345–68.

Gould, S. J. 1987. *Time's arrow, time's cycle: myth and metaphor in the discovery of geological time.* Cambridge, Massachusetts: Harvard University Press.

Granot, G. 1996. Disaster subcultures. *Disaster Prevention and Management* **5**(4), 36–40.

Green, C. H., S. M. Tunstall, M. H. Fordham 1991. The risks from flooding: which risks and whose perception? *Disasters* **15**(3), 22–36.

Greenberg, B. S. & W. Gantz (eds) 1993. *Desert Storm and the mass media.* Cresskill, New Jersey: Hampton Press.

Gruntfest, E. C. 1977. *What people did during the Big Thompson flood.* Working Paper 32, Natural Hazards Research and Applications Information Center, University of Colorado, Boulder.

Gruntfest, E. & B. Montz 1986. Changes in American urban floodplain occupancy since 1958: the experience of nine cities. *Applied Geography* **15**, 325–38.

Gruntfest, E. & M. Weber 1998. Internet and emergency management: prospects for the future. *International Journal of Mass Emergencies and Disasters* **16**(1), 55–72.

Guadagno, G. 1971. *La negazione urbana: trasformazioni sociali e comportamento deviato a Napoli.* Bologna: Il Mulino.

Guarnizo, C. C. 1993. Integrating disaster and development assistance after natural disasters: NGO response in the Third World. *International Journal of Mass Emergencies and Disasters* **11**(1), 111–22.

Guerra, A. J. T. 1995. The catastrophic events in Petropolis City (Rio de Janeiro State), between 1940 and 1990. *GeoJournal* **37**(3), 349–54.

Guettard, Jean-Etienne 1752. Mémoires sur quelques montagnes de la France qui ont été des volcans. *Mémoires de l'Académie Royale des Sciences, Paris*, 27–59.

Guidoboni, E. 1986. The earthquake of December 25, 1222: analysis of a myth. *Geologia Applicata e Idrogeologia* **21**(3), 413–24.

Guidoboni, E. & G. Ferrari (eds) 1987. *Mallet's macroseismic survey of the Neapolitan earthquake of 16th December 1857.* Bologna: Società-Geofisica-Ambiente.

Guo Huancheng, Wu Dengru, Zhu Hongxing 1989. Land restoration in China. *Journal of Applied Ecology* **26**, 787–92.

Hadfield, P. 1992. *Sixty seconds that will change the world: the coming Tokyo earthquake.* Boston: Charles E. Tuttle.

Hagen, J. 1992. *An entangled bank: the origins of ecosystem ecology.* New Brunswick, New Jersey: Rutgers University Press.

Hagman, G. 1985. *Prevention better than cure: human and environmental disasters in the Third World.* Geneva: Swedish Red Cross and League of Red Cross and Red Crescent Societies.

Hamilton, William 1772. *Observations on Mount Vesuvius, Mount Etna and other volcanoes.* London: T. Cadell.

Han Chunru 1989. Recent changes in the rural environment in China. *Journal of Applied Ecology* **26**, 803–812.

Hannigan, J. A. & R. M. Kueneman 1978. Anticipating flood emergencies: a case study of a Canadian disaster subculture. In *Disasters: theory and research*, E. L. Quarantelli (ed.), 129–36. Beverly Hills, California: Sage.

Hansson, R. O., R. J. Henze, M. A. Langenheim, A. J. Filipovitch 1979. Threat, knowledge and support for a collective response to urban flooding. *Journal of Applied Social Psychology* **9**(5), 413–25.

Hansen, G. & E. Condon 1989. *Denial of disaster.* San Francisco: Cameron.

Haque, C. E. 1988. Human adjustments to river bank erosion hazard in the Jamuna floodplain, Bangladesh. *Human Ecology* **16**(4), 421–37.

Haque, C. E. & M. Q. Zaman 1989. Coping with riverbank erosion hazard and displacement in Bangladesh: survival strategies and adjustments. *Disasters* **13**(4), 300–314.

Hardin, G. 1968. The tragedy of the commons. *Science* **162**, 1243–8.

Harvey, D. 1969. *Explanation in geography*. London: Edward Arnold.

Hassan, H. M. & W. Luscombe, W. 1991. Remote sensing and technology transfer in developing countries. In *Managing natural disasters and the environment*, A. Kreimer & M. Munasinghe (eds), 141–4. Washington DC: World Bank (Environment Department).

Hatori, T. 1996. Tsunami magnitudes in Taiwan, the Philippines and Indonesia. *GeoJournal* **38**(3), 313–19.

Hellemans, A. 1996. Ariane failure casts shadow over ESA's science program. *Science* **272**(5268), 1579.

Hendrickson, D. 1998. Humanitarian action and protracted crisis: an overview of debates and dilemmas. *Disasters* **22**(4), 283–7.

Herbich, J. B. & J. P. Haney 1982. Coastal erosion. In *The encyclopedia of beaches and coastal environments*, M. L. Schwartz (ed.), 265–7. Stroudsburg, Pennsylvania: Dowden, Hutchinson and Ross.

Hermon, E. J. & N. Chomsky 1988. *Manufacturing consent: the political economy of the mass media*. New York: Pantheon.

Hess, J. C. & J. B. Elsner 1994. Historical developments leading to current forecast models of annual Atlantic hurricane activity. *Bulletin of the American Meteorological Society* **75**(9), 1611–22.

Hewitt, K. 1970. Probablistic approaches to discrete natural events: a review and theoretical discussion. *Economic Geography* [supplement] **46**(2), 332–49.

—— 1983. The idea of calamity in a technocratic age. In *Interpretations of calamity from the viewpoint of human ecology*, K. Hewitt (ed.), 3–32. London: Allen & Unwin.

—— 1995. Excluded perspectives in the social construction of disaster. *International Journal of Mass Emergencies and Disasters* **13**(3), 317–39.

—— 1997. *Regions of risk: a geographical introduction to disasters*. Reading, Massachusetts: Addison–Wesley.

Hewitt, K. & I. Burton 1971. *The hazardousness of place*. Toronto: University of Toronto Press.

Hitchcock, R. K. & H. Hussain 1987. Agricultural and non-agricultural settlements for drought-afflicted pastoralists in Somalia. *Disasters* **11**(1), 30–39.

Hodgson, M. E. & R. Palm 1992. Attitude toward disaster: a GIS design for analyzing human response to earthquake hazards. *Geo Info Systems* (July–August), 41–51.

Hogg, S. J. 1980. Reconstruction following seismic disaster in Venzone, Friuli. *Disasters* **4**(2), 173–85.

Holland, G. L. 1989. Observations on the International Decade for Natural Disaster Reduction. *Natural Hazards* **2**(1), 77–82.

Holt-Jensen, A. 1988. *Geography: history and concepts* (2nd edn). London: Paul Chapman.

Hooke, Robert 1705. Lectures and discourses of earthquakes and subterraneous eruptions. In *The posthumous works of Robert Hooke*, 279–450. London: Richard Waller.

Hoque, M. & R. A. Khandoker 1990. Tectonics, seismicity and floods in Bangladesh. *Bangladesh Quest* **2**(1–3), 17–25.

Horlick-Jones, T. 1995a. Modern disasters as outrage and betrayal. *International Journal of Mass Emergencies and Disasters* **13**(3), 305–315.

—— 1995b. Urban disasters and megacities in a risk society. *GeoJournal* **37**(3), 329–34.

—— 1996. The problem of blame. In *Accident and design*, C. Hood & D. Jones (eds), 61–70. London: UCL Press.

Horowitz, M. 1986. *Stress response syndromes*. New York: Jason Aronson.

Hsü, K. J. 1975. Catastrophic debris streams (sturzstroms) generated by rockfalls. *Bulletin of the Geological Society of America* **86**, 129–40.

Hughes, R. 1981. Field survey techniques for estimating the normal performance of vernacular buildings prior to earthquakes. *Disasters* **5**(4), 411–15.

—— 1982. The effects of flooding upon buildings in developing countries. *Disasters* **6**(3), 183–94.

Hutton, James & John Playfair 1970. *System of the Earth, 1785, Theory of the Earth, 1788,*

Observations on granite, 1794, together with Playfair's Biography of Hutton. Darien, Connecticut: Hafner.

IFRCRCS 1994. *World disasters report 1994.* Dordrecht: Kluwer, for International Federation of Red Cross and Red Crescent societies.

—— 1996. *World disasters report 1996.* Oxford: Oxford University Press, for International Federation of Red Cross and Red Crescent societies.

—— 1997. *World disasters report 1997.* Oxford: Oxford University Press, for International Federation of Red Cross and Red Crescent societies.

—— 1998. *World disasters report 1998.* Oxford: Oxford University Press, for International Federation of Red Cross and Red Crescent societies.

Ingram, D. 1977. "Visiting firemen". *New Internationalist* **53**, 10–12.

Ippolito, F. 1985. Sulla struttura geologica dell'area napoletana. In *Napoli: una storia per immagini*, 386–90. Naples: Gaetano Macchiaroli Editore.

Irwin, R. L. 1989. The incident command system (ICS). In *Disaster response: principles of preparation and coordination*, E. Auf der Heide (ed.), 133–63. St Louis, Missouri: Mosby.

Islam, M. A. 1974. Tropical cyclones: coastal Bangladesh. In *Natural hazards, local, national, global*, G. F. White (ed.), 19–25. New York: Oxford University Press.

Isnard, G. 1758. *Mémoire sur les tremblemens de terre.* Paris.

Japanese IDNDR Committee 1995. The great Hanshin-Awaji earthquake: damages and response. *Stop Disasters* **23**, 10–13.

Jibson, R. W. 1989. The Mameyes, Puerto Rico, landslide disaster. *Landslide News* **3**, 12–14.

Jing Ke 1988. A study on the relationship between soil erosion and the geographical environment in the middle Yellow River basin. *Chinese Journal of Arid Land Research* **1**, 289–99.

Johnson, A. C. 1990. An earthquake strength scale for the media and the public. *Earthquakes and Volcanoes* **22**(5), 214–16.

Johnson, B. B. & V. T. Covello (eds) 1987. *The social and cultural construction of risk: essays on risk selection and perception.* Dordrecht: Kluwer.

Johnson, F. A. & P. Illes 1976. A classification of dam failures. *Water Power and Dam Construction* **28**, 43–5.

Johnson, N. R. 1987. Panic and the breakdown of social order: popular myth, social theory, empirical evidence. *Sociological Focus* **20**(3), 171–83.

Jones, E. L. 1987. *The European miracle: environments, economies and geopolitics in the history of Europe and Asia* (2nd edn). New York: Cambridge University Press.

Jones, F. O. 1973. *Landslides of Rio de Janeiro and the Serra das Araras escarpment, Brazil.* Professional Paper 697, US Geological Survey, Washington DC.

Jones, N. P. 1990. Considerations in the epidemiology of earthquake injuries. *Earthquake Spectra* **6**(3), 507–528.

Kakhandiki, A. & H. Shah 1998. Understanding time variation of risk: crucial implications for megacities worldwide. *Applied Geography* **18**(1), 47–53.

Kasperson, R. E. & J. X. Kasperson 1983. Determining the acceptability of risk: ethical and policy issues. In *Risk: symposium proceedings on the assessment and perception of risk to human health in Canada.* J. T. Rodgers & D. V. Bates (eds), 135–56. Ottawa: Royal Society of Canada.

Kasperson, R. E. & P. J. M. Stallen (eds) 1990. *Communicating risks to the public.* Dordrecht: Kluwer.

Kassas, M., Y. J. Ahmad, B. Rozanov 1990. Desertification and drought: an ecological and economic analysis. *Desertification Control Bulletin* **20**, 19–29.

Kates, R. W. 1962. *Hazard and choice perception in flood plain management.* Research Paper 78, Department of Geography, University of Chicago.

—— 1971. Natural hazards in human ecological perspective: hypothesis and model. *Economic Geography* **47**(3), 438–51.

Kates, R. W. & D. Pijawka 1977. From rubble to monument: the pace of reconstruction. In

Disaster and reconstruction, J. E. Haas, R. W. Kates, M. J. Bowden (eds), 1–23. Cambridge, Massachusetts: MIT Press.

Katsouyani, K., M. Kogevinas, D. Trichopoulos 1986. Earthquake-related stress and cardiac mortality. *International Journal of Epidemiology* **15**, 326–30.

Kattelmann, R. 1990. Conflicts and cooperation over the floods in the Himalaya–Ganges region. *Water International* **15**(4), 189–94.

Kaufman, W. & O. H. Pilkey 1984. *The beaches are moving*. Durham, North Carolina: Duke University Press.

Keating, J. F. 1982. The myth of panic. *Fire Journal* **77**, 57–61.

Keller, A. Z., E. L. Coles, R. Heal 1996. Experiences gained from the First Internet Disaster Conference. *Disaster Prevention and Management* **5**(5), 31-3.

Kelly, C. 1995. Assessing disaster needs in megacities: perspectives from developing countries. *GeoJournal* **37**(3), 381–5.

———1996. Further thoughts on a "disaster continuum". *Disasters* **20**(3), 276–7.

Key, D. (ed.) 1995. *Structures to withstand disasters*. London: Institution of Civil Engineers.

Khalil, G. Md. 1993. The catastrophic cyclone of April 1991: its impact on the economy of Bangladesh. *Natural Hazards* **8**(3), 263–81.

Kibreab, G. 1997. Environmental causes and impact of refugee movements: a critique of the current debate. *Disasters* **21**(1), 20–38.

Kilpatrick, F. P. 1957. Problems of perception in extreme situations. *Human Organization* **16**(2), 20–22.

Kircher, Athanasius 1664–5. *Mundus subterraneus in XII libros digestus: quo divinum subterrestris mundi opificium* (2 vols). Amsterdam.

Kircher, C. A. & P. Stojanovski 1996. An overview of the "HAZUS" earthquake loss estimation methodology. In *Second National Workshop on Modeling Earthquake Casualties for Planning and Response: summary of proceedings*, R. A. Olson & D. E. Alexander (eds), 1–3. Sacramento, California: Robert Olson Associates.

Kirkby, J., P. O'Keefe, I. Convery, D. Howell 1997. On the emergence of complex disasters. *Disasters* **21**(2), 177–80.

Kisslinger, C. 1974. Earthquake prediction studies in China. *Physics Today* **27**(3), 36–42.

Knight, C. 1990. Un po' per diletto, e un po' per guadagnare: Sir William Hamilton e il mondo della committenza straniera a Napoli. In *All'ombra del Vesuvio: Napoli nella veduta europea dal Quattrocento all'Ottocento*, 95–107. Naples: Electa Napoli.

Kreps, G. A. 1983. The organization of disaster response: core concepts and processes. *International Journal of Mass Emergencies and Disasters* **1**(3), 439–66.

———1989. Description, taxonomy and explanation in disaster research. *International Journal of Mass Emergencies and Disasters* **7**(3), 277–80.

———1995. Disaster as systemic event and social catalyst: a clarification of subject matter. *International Journal of Mass Emergencies and Disasters* **13**(3), 255–84.

Krimgold, F. 1989. Search and rescue in the 1988 earthquake in Soviet Armenia. *Earthquake Spectra* [special supplement], 136–49.

Kroll-Smith, J. & S. R. Couch 1991. What is disaster? An ecological–symbolic approach to resolving the definitional debate. *International Journal of Mass Emergencies and Disasters* **9**(3), 355–66.

Kuhn, T. S. 1962. *The structure of scientific revolutions*. Chicago: University of Chicago Press.

Kundzewicz, Z. W., et al. (eds) 1993. *Extreme hydrological events: precipitation, floods, and droughts*. Publication 213, International Association of Hydrological Sciences, Washington DC.

Kunreuther, H. 1974. Economic analysis of natural hazards: an ordered choice approach. In *Natural hazards: local, national, global*, G. F. White (ed.), 206–214. New York: Oxford University Press.

Kunreuther, H. & R. J. Roth Sr (eds) 1998. *Paying the price: the status and role of insurance against natural disasters in the United States*. Washington DC: National Academy Press.

Kusler, J. A. 1985. Liability as a dilemma for local managers. *Public Administration Review* **45** [special issue], 118–22.

Larsson, G. & A. Enander 1997. Preparing for disaster: public attitudes and actions. *Disaster Prevention and Management* **6**(1), 11–21.

Latif, A. 1969. Investigation of the Brahmaputra River. *Proceedings of the American Society of Civil Engineers, Journal of the Hydraulics Division* **95**(HY5), 1687–98.

Lavell, A. 1994. Opening a policy window: the Costa Rican Hospital Retrofit and Seismic Insurance Programs 1986–1992. *International Journal of Mass Emergencies and Disasters* **12**(1), 95–115.

Leader, N. 1998. Proliferating principles; or, how to sup with the Devil without getting eaten. *Disasters* **22**(4), 288–308.

Leighton, F. B. 1976. Urban landslides: targets for land-use planning in California. In *Urban geomorphology*, D. R. Coates (ed.), 37–60. Special Paper 174, Geological Society of America, Boulder, Colorado.

Leitko, T. A., D. R. Rudy, S. A. Peterson 1980. Loss not need: the ethics of relief giving in natural disasters. *Journal of Sociology and Social Welfare* **7**, 730–41.

Leonards, G. A. (ed.) 1987. Dam failures. *Engineering Geology* **24**(1–4), 1–612.

Leone, U. 1991. *Dalla memoria allo sguardo: viaggio nella qualità della vita a Napoli*. Naples: CUEN.

Leor, J. & R. A. Kloner 1996. The Northridge earthquake as a trigger for acute myocardial infarction. *American Journal of Cardiology* **77**(14), 1230–32.

Levi, L. 1997. Does number of beds reflect the surgical capability of hospitals in wartime and disaster? The use of a simulation technique at a national level. *Prehospital and Disaster Medicine* **12**, 300–310.

Lindell, M. K. 1994. Perceived characteristics of environmental hazards. *International Journal of Mass Emergencies and Disasters* **12**(3), 303–326.

Lindell, M. K. & R. W. Perry 1992. *Behavioral foundations of community emergency planning*. Bristol, Pennsylvania: Taylor & Francis.

Lindvall, C. M. & A. J. Nitko 1975. *Measuring pupil achievement and aptitude* (2nd edn). New York: Harcourt Brace Jovanovich.

Lirer, L., G. Luongo, R. Scandone 1987. On the volcanological evolution of Campi Flegrei. EOS: *Transactions of the American Geophysical Union* **68**, 226–34.

Lockridge, P. A. 1988. Historical tsunamis in the Pacific basin. In *Natural and man-made hazards*, M. I. El-Sabh & T. S. Murty (eds), 171–81. Dordrecht: Reidel.

Loescher, G. I. & L. Monahan (eds) 1990. *Refugees and international relations*. Oxford: Oxford University Press.

Lord Judd of Portsea 1992. Disaster relief or relief disaster? A challenge to the international community. *Disasters* **16**(1), 1–8.

Luca, M. 1969. *Gli economisti napoletani del settecento e la politica dello sviluppo*. Naples: Morano.

Lu Jingshen, Du Gangjian, Song Gang 1992. The experience, lesson and reform of China's disaster management. *International Journal of Mass Emergencies and Disasters* **10**(2), 315–27.

Lumb, P. 1975. Slope failures in Hong Kong. *Quarterly Journal of Engineering Geology* **8**(1), 31–65.

Lyell, Charles 1830–33. *Principles of geology* (3 vols). London: John Murray [Reissued 1990–91, edited by M. Rudwick. Chicago: University of Chicago Press.].

Mack, P. E. 1990. *Viewing the Earth: the social construction of the Landsat satellite system*. Cambridge, Massachusetts: MIT Press.

Macrae, J. 1998. The death of humanitarianism? An anatomy of the attack. *Disasters* **22**(4), 309–317.

Macrae, J., M. Bradbury, S. Jaspars, D. Johnson, M. Duffield 1997. Conflict, the continuum and chronic emergencies: a critical analysis of the scope for linking relief, rehabilitation and development planning in Sudan. *Disasters* **21**(3), 223–43.

Macrae, J. & A. B. Zwi 1992. Food as an instrument of war in contemporary African famines: a review of the evidence. *Disasters* **16**(4), 299–321.

Macry, P. 1974. *Mercato e società nel Regno di Napoli*. Naples: Guida Editori.

Mahaney, W. C. (ed.) 1984. *Quaternary dating methods*. Amsterdam: Elsevier.

Mamun, M. Z. 1996. Awareness, preparedness and adjustment measures of river-bank erosion-prone people. *Disasters* **20**(1), 68–74.

Maravall, J. A. 1979. La cultura de crisis barocca. *Historia* **16**, 80–90.

Marmar, C. R., D. S. Weiss, T. J. Metzler, K. Delucchi 1996. Characteristics of emergency services personnel related to peritraumatic dissociation during critical incident exposure. *American Journal of Psychiatry* **153**(7, supplement), 94–102.

Masellis, M. & S. W. A. Gunn (eds) 1992. *The management of mass burn casualties and fire disasters.* Dordrecht: Kluwer.

Maskrey, A. 1989. *Disaster mitigation: a community based approach.* Oxford: Oxfam.

Mastriani, R. 1921. *Cenni storico-cartografici sul Vesuvio.* Florence: L'Universo.

Matsuda, I. 1996. Two surveys for taking measures to cope with the coming earthquake to the Tokyo metropolis. *GeoJournal* **38**(3), 349–53.

May, P. J. 1985. *Recovering from catastrophes: federal disaster relief policy and politics.* Westport, Connecticut: Greenwood Press.

Mayer, J. 1969. Famine in Biafra. *Post-Graduate Medicine* **45**, 236–40.

Mayer, L. & D. Nash (eds) 1987. *Catastrophic flooding.* London: Allen & Unwin.

McCain, J. F., L. R. Hoxit, R. A. Maddox, C. F. Chappell, F. Caracena 1979. *Storm and flood of July 31 – August 1, 1976, in the Big Thompson River and Cache la Poudre River basins, Larimer and Weld Counties, Colorado.* Professional Paper 1115, US Geological Survey, Washington DC.

McGill, W. L. 1957. How a state prepares for disaster. *Annals of the American Academy of Political and Social Science* **309**, 89–97.

McGuire, W. J., C. R. J. Kilburn, J. B. Murray (eds) 1994. *Monitoring active volcanoes.* London: UCL Press.

McKay, J. M. 1983. Newspaper reporting of bushfire disaster in southeastern Australia: Ash Wednesday 1983. *Disasters* **7**(4), 283–90.

McKee, C. O., P. L. Lowenstein, P. de St Ours, B. Talai, I. Itikarai, J. J. Mori 1984. Seismic and ground deformation crises at Rabaul caldera, prelude to an eruption? *Bulletin Volcanologique* **47**, 397–411.

Meadows, D. H., et al. 1972. *The limits to growth: a report for the Club of Rome's project on the predicament of mankind.* New York: Universe Books.

Meyer-Abich, K. M. 1997. Humans in nature: toward a physiocentric philosophy. In *Technological trajectories and the human environment*, J. H. Ausubel & H. D. Langford (eds), 168–84. Washington DC: National Academy of Engineering and National Academy Press.

Middleton, N., P. O'Keefe, S. Mayo 1993. *The tears of the crocodile: from Rio to reality in the developing world.* London: Pluto Press.

Mileti, D. S. 1980. Human adjustment to the risk of environmental extremes. *Sociology and Social Research* **64**(3), 328–47.

Miller, J. 1974. *Aberfan: a disaster and its aftermath.* London: Constable.

Mitchell, J. 1983. When disaster strikes: the critical incident stress debriefing process. *Journal of Emergency Medical Services* **8**, 36–9.

Mitchell, J. K. 1984. Hazard perception studies: convergent concerns and divergent approaches during the last decade. In *Environmental perception and behavior: an inventory and prospect*, T. F. Saarinen, D. Seamon, J. L. Sell (eds), 33–59. Research Paper 209, Department of Geography, University of Chicago.

—— 1993. Natural hazard predictions and responses in very large cities. In *Prediction and perception of natural hazards*, J. Nemec, J. M. Nigg, F. Siccardi (eds), 29–37. Dordrecht: Kluwer.

—— 1995a. Coping with natural hazards and disasters in megacities: perspectives on the twenty-first century. *GeoJournal* **37**(3), 303–311.

—— (ed.) 1995b. *Megacities and natural disasters.* Tokyo: United Nations University Press.

—— 1998. Introduction: hazards in changing cities. *Applied Geography* **18**(1), 1–6.

Mitchell, J. K., N. Devine, K. Jagger 1989. A contextual model of natural hazard. *Geographical Review* **79**(4), 391–409.

Mitchell, J. T. 1997. Can hazard risk be communicated through a virtual experience? *Disasters* **21**(3), 258–66.

Monmonier, M. 1997. *Cartographies of danger: mapping hazards in America*. Chicago: University of Chicago Press.

Moore, P. S., et al. 1993. Mortality rates in displaced and resident populations of Central Somalia during the famine of 1992. *The Lancet* **341**, 935–8.

Moran, C., N. R. Britton, B. Correy 1992. Characterising voluntary emergency responders. *International Journal of Mass Emergencies and Disasters* **10**(1), 207–216.

Morherg-Schulz, C. 1980. *Genius loci: towards a phenomenology of architecture*. New York: Rizzoli.

Mould, R. F. 1992. *Chernobyl: the real story*. Oxford: Pergamon.

Museo Pignatelli 1985. *Goaches napoletane del Settecento e dell'Ottocento*. Naples: Electa.

Mushtaque, A., R. Chowdhury, A. U. Bhuyia, A. Y. Chowdhury, R. Sen 1993. The Bangladesh cyclone of 1991: why so many people died. *Disasters* **17**(4), 291–303.

Myers, M. F. & G. F. White 1993. The challenge of the Mississippi floods. *Environment* **35**(10), 6–9, 25–35.

Natsios, A. S. 1991. Economic incentives and disaster mitigation. In *Managing natural disasters and the environment*, A. Kreimer & M. Munasinghe (eds), 111–14. Washington DC: World Bank (Environment Department).

Neal, D. M. 1984. Blame assignment in a diffuse disaster situation: a case example of the role of an emergent citizen group. *International Journal of Mass Emergencies and Disasters* **2**(2), 251–66.

——— 1997. Reconsidering the phases of disaster. *International Journal of Mass Emergencies and Disasters* **15**(2), 239–64.

Needham, R. D. 1986. The cosmopolite–localite model: newspaper types of natural hazard information potentials. *Environmental Management* **10**, 271–84.

Nemec, J., J. M. Nigg, F. Siccardi (eds) 1993. *Prediction and perception of natural hazards*. Dordrecht: Kluwer.

Newman, M. C. & C. L. Strojan 1998. *Risk assessment: logic and measurement*. Chelsea, Michigan: Ann Arbor Press.

NIBS 1997. *HAZUS: user's manual and technical manuals*, vols 1–3 (4 vols). Washington DC: National Institute of Building Sciences and Federal Emergency Management Agency.

Noji, E. K. (ed.) 1997. *The public health consequences of disasters*. New York: Oxford University Press.

Nolen-Hoekesema, S. & J. Morrow 1991. A prospective study of depression and post-traumatic stress symptoms after a natural disaster: the 1989 Loma Prieta earthquake. *Journal of Personality and Social Psychology* **62** (3), 115–31.

Normile, D. 1994. Japan holds firm to shaky science. *Science* **264**, 1656–8.

Nurul Alam, S. M. 1990. Perceptions of flood among Bangladeshi villagers. *Disasters* **14**(4), 354–8.

O'Brien, P. W. & D. S. Mileti 1992. Citizen participation in emergency response following the Loma Prieta earthquake. *International Journal of Mass Emergencies and Disasters* **10**(1), 71–89.

Odemerho, F. 1993. Flood control failures in a Third World city: Benin City, Nigeria, some environmental factors and policy issues. *GeoJournal* **29**(4), 371–6.

Okimura, T., S. Takada, T. H. Koid 1996. Outline of the Great Hanshin Earthquake, Japan, 1995. *Natural Hazards* **14**(1), 39–71.

Okrent, D. 1980. Comment on societal risk. *Science* **208**, 372–5.

Oldroyd, D. 1972. Robert Hooke's methodology of science as exemplified by his discourse on earthquakes. *British Journal for the History of Science* **6**, 109–30.

Oliver-Smith, A. 1977a. Disaster rehabilitation and social change in Yungay, Perú. *Human Organization* **36**, 5–13.

——— 1977b. Traditional agriculture, central places and post-disaster urban relocation in Peru. *American Ethnologist* **4**(1), 102–116.

——— 1986. *The martyred city: death and rebirth in the Andes*. Albuquerque: University of New Mexico Press.

——— 1991. Successes and failures in post-disaster resettlement. *Disasters* **15**(1), 12–23.

Olser, T. 1993. Injury severity scoring: perspectives in development and future directions. *American Journal of Surgery* **165**(2A, supplement), 43S–51S.

Olson, R. A. & D. E. Alexander (eds) 1996. *Second National Workshop on Modeling Earthquake Casualties for Planning and Response: summary of proceedings*. Sacramento, California: Robert Olson Associates.

Olson, R. A. & R. S. Olson 1977. The Guatemala earthquake of 4 February 1976: social science observations and research suggestions. *Mass Emergencies* **2**, 69–81.

Olson, R. S. 1995. Guns, drugs, and disaster: Cauca/Huila, Colombia, 1994. *International Journal of Mass Emergencies and Disasters* **13**(2), 147–60.

Olson, R. S. & R. A. Olson 1987. Urban heavy rescue. *Earthquake Spectra* **3**(4), 645–58.

Omaar, R. & A. de Waal 1992. Disaster pornography from Somalia. *Los Angeles Times* (8 June).

Otway, H. & K. Thomas 1982. Reflections in risk perception and policy. *Risk Analysis* **2**(2), 69–82.

Ouellette, P., D. Leblanc, N. El-Jabi, J. Rouselle 1988. Cost–benefit analysis of flood-plain zoning. *Journal of Water Resources Planning and Management* **114**(3), 326–34.

Packard, V. O. 1981. *The hidden persuaders*. New York: Pocket Books.

Palm, R. 1998. Urban earthquake hazards: the impacts of culture on perceived risk and response in the USA and Japan. *Applied Geography* **18**(1), 35–46.

Pane, R., G. Alisio, P. Di Monda, L. Santoro, A. Venditti 1959. *Ville vesuviane del settecento*. Naples: Edizioni Scientifiche Italiane.

Parker, D. J. & D. M. Harding 1979. Natural hazard evaluation, perception and adjustment. *Geography* **64**(4), 307–316.

Parr, A. R. 1970. Organizational response to community crises and group emergence. *American Behavoural Scientist* **13**, 423–9.

Paul, B. K. 1984. Perception and agricultural adjustments to floods in the Jamuna floodplain, Bangladesh. *Human Ecology* **12**(1), 3–19.

——— 1997. Flood research in Bangladesh in retrospect and prospect: a review. *Geoforum* **28**(2), 121–32.

——— 1999. Women's awareness of and attitudes towards the Flood Action Plan (FAP) in Bangladesh: a comparative study. *Environmental Management* **23**(1), 103–114.

Pepper, D. 1984. *The roots of modern environmentalism*. London: Croom–Helm.

——— 1996. *Modern environmentalism: an introduction*. London: Routledge.

Perry, R. W. 1985. *Comprehensive emergency management*. Greenwich, Connecticut: JAI Press.

Perry, R. W., D. F. Gillespie, D. S. Mileti 1974. System stress and the persistence of emergent organizations. *Sociological Inquiry* **44**(2), 111–19.

Pestel, E. (ed.) 1989. *Beyond the limits to growth: a report to the Club of Rome*. New York: Universe Books.

Petak, W. J. & A. A. Atkisson 1982. *Natural hazard risk assessment and public policy: anticipating the unexpected*. New York: Springer.

Petraccone, C. 1975. *Napoli dal Cinquecento all'Ottocento: problemi di storia demografica e sociale*. Naples: Guida Editori.

——— 1981. *Napoli moderna e contemporanea*. Naples: Guida Editori.

Petrov V. A., A. O. Mostryukov, V. I. Lykov 1994. The recent field of tectonic stresses over the territory of China. *Journal of Earthquake Prediction Research* **3**(4), 509–527.

Phillips, J. 1865. *Vesuvius*. Oxford: Oxford University Press.

Phillips, J. D. 1993. Biophysical feedbacks and the risks of desertification. *Annals of the Association of American Geographers* **83**(4), 630–40.

Pielke, R. A. 1995. *Hurricane Andrew in south Florida: mesoscale weather and societal responses*. National Center for Atmospheric Research, Boulder, Colorado.

Pierson, T. C. 1992. Rainfall triggered lahars at Mount Pinatubo, Philippines, following the June 1991 eruption. *Landslide News* **6**, 6–9.

Pinatubo Volcano Observatory Team 1991. Lessons from a major eruption: Mount Pinatubo,

Philippines. EOS: Transactions of the American Geophysical Union **72**, 545–55.

Pineda, A. L. 1993. The Philippines: common haven of natural disasters. INCEDE Newsletter **2**(3), 3–8.

Placanica, A. 1985. *Il filosofo e la catastrofe: un terremoto del settecento*. Turin: Einaudi.

Plafker, G., G. E. Ericksen, J. Fernández Concha 1971. Geological aspects of the May 31, 1970, Peru earthquake. *Bulletin of the Seismological Society of America* **61**(3), 543–78.

Plato 1962. *Timæus, Critias, Cleitophan, Menexemus and Epistles* (translated by R. G. Bury). Cambridge, Massachusetts: Harvard University Press.

Pliny, C. 1956. *Natural history* (translated by H. Rackham). Cambridge, Massachusetts: Harvard University Press.

Ploughman, P. 1995. The American print news media "construction" of five natural disasters. *Disasters* **19**(4), 308–326.

—— 1997. Disasters, the media and social structures: a typology of credibility hierarchy persistence based on newspaper coverage of the Love Canal and six other disasters. *Disasters* **21**(2), 118–37.

Porfiriev, B. N. 1996. Social aftermath and organizational response to a major disaster: the case of the 1995 Sakhalin earthquake in Russia. *Journal of Contingencies and Crisis Management* **4**(4), 218–27.

—— 1997. *Disaster policy and emergency management in Russia*. New York: Nova Science Publishers.

Pramanik, M. A. H. 1995. *Disaster mitigation strategies in Bangladesh*. INCEDE Report 1995-01, International Center for Disaster Mitigation Engineering, University of Tokyo.

Preston, V., S. M. Taylor, D. C. Hodge 1983. Adjustment to natural and technological hazards: a study of an urban residential community. *Environment and Behaviour* **15**(2), 143–64.

Pugh, M. 1998. Military intervention and humanitarian action: trends and issues. *Disasters* **22**(4), 339–51.

Pyle, A. S. 1992. The resilience of households to famine in El Fashar, Sudan, 1982–89. *Disasters* **16**(1), 19–27.

Pyle, D. M. 1997. The global impact of the Minoan eruption of Santorini, Greece. *Environmental Geology* **30**(1–2), 59–61.

Quarantelli, E. L. 1954. The nature and conditions of panic. *American Journal of Sociology* **60**(3), 267–75.

—— (ed.) 1978. *Disasters: theory and research*. Beverly Hills, California: Sage.

—— 1989. Conceptualizing disasters from a sociological perspective. *International Journal of Mass Emergencies and Disasters* **7**(3), 243–52.

—— 1995. What is disaster? Six views of the problem. *International Journal of Mass Emergencies and Disasters* **13**(3), 221–364.

—— 1996. Problematical aspects of the information/communication revolution for disaster planning and disaster research: ten non-technical issues and questions. "Disaster 96: Electronic Communication and Disaster Management", First Internet Conference, Bradford, England [Internet publication].

—— 1997. The Disaster Research Center field studies of organized behavior in the crisis time period of disasters. *International Journal of Mass Emergencies and Disasters* **15**(1), 47–70.

—— (ed.) 1998. *What is a disaster? Perspectives on the question*. London: Routledge.

Rantz, S. E. 1970. *Urban sprawl and flooding in southern California*. Circular 601B, US Geological Survey, Washington DC.

Rappaport, E. N. 1994. Hurricane Andrew. *Weather* **49**, 51–61.

Rasid, H. 1982. Urban flood problem in Benin City, Nigeria: natural or man-made? *Malaysian Journal of Tropical Geography* **6**, 17–30.

Rasid, H., D. Baker, R. Kreutzwiser 1992. Coping with Great Lakes flood and erosion hazards:

Long Point, Lake Erie, vs Minnesota Point, Lake Superior. *Journal of Great Lakes Research* **18**(1), 29–42.

Rasid, H. & A. Mallik 1993. Poldering versus compartmentalization: the choice of flood control techniques in Bangladesh. *Environmental Management* **17**(1), 59–71.

Rasid, H. & B. K. Paul 1987. Flood problems in Bangladesh: is there an indigenous solution? *Environmental Management* **11**(2), 155–73.

Rasid, H. & M. A. H. Pramanik 1990. Visual interpretation of satellite imagery for monitoring floods in Bangladesh. *Environmental Management* **14**(6), 815–21.

—— 1993. Areal extent of the 1988 flood in Bangladesh: how much did the satellite imagery show? *Natural Hazards* **8**(2), 189–200.

Ray, John 1968(1692). *Miscellaneous discourses concerning the dissolution and changing of the world*. Hildesheim: G. Olms.

Reitherman, R. 1985. A review of earthquake damage estimation methods. *Earthquake Spectra* **1**(4), 805–849.

Rendell, H. M. 1985. Earthquake damage in Tricarico, southern Italy. *Environmental Management* **9**(4), 289–302.

Rendell, H. M. & D. E. Alexander 1985. Tricarico: il dopoterremoto. *Bollettino della Biblioteca Provinciale di Matera* **6**, 47–57.

Righi, P. V., G. Marchi, G. Dondi 1986. Stabilizzazione mediante pozzi drenanti di un movimento franoso nella città di Perugia. *Associazione Geotecnica Italiana, 16th National Conference on Engineering Geology, Bologna 14–16 May 1986*: 167–77. Bologna: Associazione Geotecnica Italiana.

Rikitake, T. (ed.) 1984. *Earthquake prediction*. Tokyo and Paris: Terra Scientific Publishing and UNESCO Press.

—— 1990. Threat of an earthquake right under the capital of Japan. *Earthquakes and Volcanoes* **22**(5), 209–210.

RMS 1993. *Assessment of the state-of-the-art earthquake loss estimation methodologies* [promotional brochure]. Washington DC: Risk Management Software Inc. and National Institute of Building Sciences.

Robinson, A. R. 1981. Erosion and sediment control in China's Yellow River Basin. *Journal of Soil and Water Conservation* **36**, 125–7.

Rosenthal, U. & A. Kouzmin 1997. Crises and crisis management: toward comprehensive government decision making. *Journal of Public Administration Research and Theory* **7**(2), 277–305.

Roth, R. 1970. Cross-cultural perspectives in disaster response. *American Behavioural Scientist* **13**(3), 440–51.

Royal Society 1998. *Preventing natural disasters: the role of risk control and insurance*. London: Royal Society.

Russell, C. S. 1970. Losses from natural hazards. *Land Economics* **46**, 383–93.

Ryan, M. P. (ed.) 1990. *Magma storage and ascent*. Chichester, England: John Wiley.

Ryle, J. 1992. Notes on the repatriation of Somali refugees from Ethiopia. *Disasters* **16**(2), 160–68.

Saarinen, T. F. 1966. *Perception of drought hazard on the Great Plains*. Research Paper 106, Department of Geography, University of Chicago, Chicago.

Sagan, C. & R. Turco 1991. *A path no man thought: nuclear winter and the end of the arms race*. New York: Random House.

Sapir, D. G. 1993. Health effects of earthquakes and volcanoes: epidemiological and policy issues. *Disasters* **17**(3), 255–62.

Sapir, D. G. & C. Misson 1992. The development of a database on disasters. *Disasters* **16**(1), 74–80.

Saunders, S. L. & G. Kreps 1987. The life history of the emergent organization in times of disaster. *Journal of Applied Behavioral Science* **23**(4), 443–62.

Scandone, R. 1977. Il rischio da colate di lava e implicazioni socio-economiche. In *Proceedings of "I vulcani attivi dell'area napoletana"*, 103–106, Atti del Convegno, Regione Campania,

Naples, June 1977.

Scanlon, J. 1988. Winners and losers: some thoughts about the political economy of disaster. *International Journal of Mass Emergencies and Disasters* **6**(1), 47–64.

—— 1992. *Convergence revisited: a new perspective on a little-studied topic*. Working Paper 75, Natural Hazards Research and Applications Information Center, University of Colorado, Boulder.

—— 1994. The role of EOCs in emergency management: a comparison of Canadian and American experience. *International Journal of Mass Emergencies and Disasters* **12**(1), 51–75.

Scanlon, J., S. Alldred, A. Farrell, A. Prawzick 1985. Coping with the media in disasters: some predictable problems. *Public Administration Review* **45**(special issue), 123–33.

Schneider, S. 1992. Governmental responses to disasters: the conflict between bureaucratic procedures and emergent norms. *Public Administration Review* **52**(2), 135–43.

Scholz, C. 1997. Whatever happened to earthquake prediction? *Geotimes* **42**(3), 16–19.

Schware, R. 1982. Official and folk flood warning systems: an assessment. *Environmental Management* **6**(3), 209–216.

Seed, H. B. & S. D. Wilson 1967. The Turnagain Heights landslide, Anchorage. *Proceedings of the American Society of Civil Engineers, Journal of the Soil Mechanics and Foundations Division* **93**(SM4), 325–53.

Seneca 1971–2. *Natural questions* (translated by T. H. Corcoran; 2 vols). Cambridge, Massachusetts: Harvard University Press.

Seydlitz, R., J. W. Spencer, G. Lundskow 1994. Media presentations of a hazard event and the public's response: an empirical examination. *International Journal of Mass Emergencies and Disasters* **12**(3), 279–301.

Shah, H. C. 1995. Scientific profiles of the "Big One". *Disaster Research* **179** [Internet bulletin] and *Natural Hazards Observer* (November 1995), Natural Hazards Center, University of Colorado, Boulder.

Shahabuddin, Q. & S. Mestelman 1986. Uncertainty and disaster avoidance behaviour in peasant farming: evidence from Bangladesh. *Journal of Development Studies* **22**(4), 740–52.

Shea, J. H. 1983. Creationism, uniformitarianism, geology, and science. *Journal of Geological Education* **31**, 105–110.

Shore, J. H. (ed.) 1986. *Disaster stress studies: new methods and findings*. Washington DC: American Psychiatric Press.

Short, J. F. O. & E. A. Rosa 1998. Organizations, disasters, risk analysis and risk: historical and contemporary contexts. *Journal of Contingencies and Crisis Management* **6**(2), 93–6.

Shrader-Frechette, K. S. 1990. Perceived risks versus actual risks: managing hazards through negotiation. *Risk: Health, Safety and Environment* **1**(4), 341–52.

Sigurdsson, H., S. Carey, W. Cornell, T. Pescatore 1985. The eruption of Vesuvius in AD 79. *National Geographic Research* **1**, 332–87.

Silver, P. & H. Wakita 1996. A search for earthquake precursors. *Science* **273**, 77–8.

Simon, H. A. 1956. Rational choice and the structure of the environment. *Psychological Review* **63**, 129–38.

Simpson-Housley, P. & P. Bradshaw 1978. Personality and the perception of earthquake hazard. *Australian Geographical Studies* **16**, 65–72.

Sims, J. H. & D. Baumann 1972. The tornado threat: coping styles of North and South. *Science* **176**, 1386–92.

Singer, E. & P. Endreny 1993. *Reporting on risk: how the mass media portray accidents, diseases, disasters and other hazards*. New York: Russell Sage Foundation.

Slim, H. 1997. Doing the right thing: relief agencies, moral dilemmas and moral responsibility in political emergencies and war. *Disasters* **21**(3), 244–57.

Slovic, P., B. Fischhoff, S. Lichtenstein 1979. Rating the risks. *Environment* **21**(3), 14–17, 36–9.

Smith, K. 1996. *Environmental hazards: assessing risk and reducing disaster* (2nd edn). London: Routledge.

Smith, R. R. 1979. Mitigation, risk aversion and regional differentiation. *Journal of Regional Science* **19**(1), 31–45.

Snarr, D. N. & E. L. Brown 1994. Post-disaster housing reconstruction: a longitudinal study of resident satisfaction. *Disasters* **18**(1), 76–80.

Solecki, W. D. & S. Michaels 1994. Looking through the post-disaster policy window. *Environmental Management* **18**(4), 587–95.

Solow, R. M. 1996. Intergenerational equity, yes – but what about inequity today? In *Human development report* 16. Oxford: Oxford University Press for United Nations Development Programme.

Song, K. L., X. Feng, J. D. Wang, J. T. Wang 1989. Landslide and debris flow disasters in China. *Landslide News* **3**, 20–21.

Sorensen, J. H. & G. F. White 1980. Natural hazards: a cross-cultural perspective. In *Environment and culture*, I. Altmann, A. Rapoport, J. F. Wohlwill (eds), 279–318. New York: Plenum Press.

Sorkin, A. L. 1983. *Economic aspects of natural hazards.* Lexington, Massachusetts: Lexington Books.

Southworth, C. S. 1985. *Characteristics and availability of data from Earth-imaging satellites.* Bulletin 1631, US Geological Survey, Washington DC.

Spooner, D. J. 1984. The Southern Problem: the Neapolitan problem and Italian regional policy. *Geographical Journal* **150**(1), 11–26.

Spurgeon, T. 1996. Concepts and Guidelines for Emergency Managers Preparing Emergency Communications Plans. Online document: http://hoshi.cic.sfu.ca/tc/contpage.htm.

Starr, C. 1969. Societal benefit versus technological risk. *Science* **165**, 1232–8.

Steedman, S. 1995. Megacities: the unacceptable risk of natural disaster. *Built Environment* **21**(2/3), 89–93.

Stephenson, R. S. 1981. *Understanding earthquake relief: guidelines for private agencies and commercial organizations.* London: International Disaster Institute.

Stephenson, R. & P. S. Anderson 1997. Disasters and the information technology revolution. *Disasters* **21**(4), 305–334.

Stillitani, E., A. Bottari, A. Teramo 1995. On macroseismic attenuation parameters in the comparison of theoretical seismic hazard maps for southern Calabria and eastern Sicily (southern Italy). *Natural Hazards* **11**(3), 259–82.

Stockton, N. 1996. Defensive development? Re-examining the role of the military in complex political emergencies. *Disasters* **20**(2), 144–8.

—— 1998. In defence of humanitarianism. *Disasters* **22**(4), 352–60.

Stop Disasters 1996. RIBAMOD: flood management in Europe. *Stop Disasters* **29**, 27.

Strabo 1969. *The geography* (translated by H. L. Jones & R. S. Sterrett). Cambridge, Massachusetts: Harvard University Press.

Strazzullo, F. 1968. *Edilizia e urbanistica a Napoli dal '500 al '700.* Naples: Arturo Berisio Editore.

Stukeley, William 1750. *The philosophy of earthquakes, natural and religious.* London: C. Corbet.

Susman, P., P. O'Keefe, B. Wisner 1983. Global disasters: a radical interpretation. In *Interpretations of calamity*, K. Hewitt (ed.), 263–83. London: Allen & Unwin.

Svoboda, J. 1999. Homosphere. In *The encyclopedia of environmental science*, D. E. Alexander & R. W. Fairbridge (eds), 324–5. Dordrecht: Kluwer.

Svoboda, J. & D. Nabert 1999. Noosphere. In *The encyclopedia of environmental science*, D. E. Alexander & R. W. Fairbridge (eds), 438–9. Dordrecht: Kluwer.

Sweet, S. 1998. The effect of a natural disaster on social cohesion: a longitudinal study. *International Journal of Mass Emergencies and Disasters* **16**(3), 321–32.

Sylves, R. T. 1991. Adopting integrated emergency management in the United States: political and cultural challenges. *International Journal of Mass Emergencies and Disasters* **9**(3), 413–24.

Tayag, J. C. & R. S. Punongbayam 1994. Volcanic disaster mitigation in the Philippines: experience from Mount Pinatubo. *Disasters* **18**(1), 1–15.

Taylor, A. J. W. 1984. Architecture and society: disaster structures and human stress. *Ekistics* **51**, 446–51.

Thatcher, W. 1976. Episodic strain accumulation in southern California. *Science* **194**, 691–6.

Tilling, R. I. 1989. Volcanic hazards and their mitigation: progress and problems. *Reviews of Geophysics* **27**(2), 237–69.

Tinkler, K. J. 1985. *A short history of geomorphology*. Totowa, New Jersey: Barnes & Noble.

Tobin, G. A. & B. E. Montz 1997. *Natural hazards: explanation and integration*. New York: Guilford Press.

Tobriner, S. 1982. *The genesis of Noto*. Berkeley: University of California Press.

Toole, M. J. & R. J. Waldman 1991. Nowhere a promised land: the plight of the world's refugees. *Encyclopedia Britannica, Medical Health Annual*, 124–41.

Torry, W. I. 1979. Hazards, hazes and holes: a critique of *The environment as hazard* and general reflections on disaster research. *Canadian Geographer* **23**(4), 368–83.

Townshend, J. R. G. 1981. The spatial resolving power of Earth resources satellites. *Progress in Physical Geography* **5**, 32–55.

Trichopoulos, D., et al. 1983. Psychological stress and fatal heart attack: the Athens (1981) earthquake natural experiment. *The Lancet* **I**(8322), 441–4.

Tubbesing, S. K. & D. S. Mileti (ed.) 1994. *The Loma Prieta, California, earthquake of October 17, 1989: loss estimation procedures societal response*. Professional Paper 1553A, US Geological Survey, Washington DC.

Turner, B. A. 1976. The development of disasters: a sequence model for the analysis of the origin of disasters. *Sociological Review* **24**(4), 753–74.

—— 1979. The social aetiology of disasters. *Disasters* **3**(1), 53–9.

Turner, R. H., J. M. Nigg, D. Heller Paz 1986. *Waiting for disaster: earthquake watch in California*. Berkeley: University of California Press.

Turton, D. 1991. Warfare, vulnerability and survival: a case from southwestern Ethiopia. *Disasters* **15**(3), 254–64.

Twigg, J. & M. R. Bhatt 1998. *Understanding vulnerability: South Asian perspectives*. London: Intermediate Technology Publications.

Uitto, J. I. 1998. The geography of disaster vulnerability in megacities. *Applied Geography* **18**(1), 7–16.

UNDP 1996. *United Nations Human Development Report 1996*. Oxford: Oxford University Press.

UNDRO 1982. *Natural disasters and vulnerability analysis*. Geneva: Office of the United Nations Disaster Relief Coordinator (UNDRO).

—— 1984. *Disaster prevention and mitigation: a compendium of current knowledge*, vol. 11: *preparedness aspects*. Geneva: Office of the United Nations Disaster Relief Coordinator (UNDRO).

US National Research Council 1987. *Confronting natural disasters: an international decade for natural hazard reduction*. Washington DC: National Academy Press.

—— 1990. *Managing coastal erosion*. Washington DC: National Academy Press.

—— 1992. *The economic consequences of a catastrophic earthquake*. Washington DC: National Academy Press for Committee on Earthquake Engineering and National Research Council.

Van Ardsol, M. G., J. Sabagh, F. Alexander 1964. Reality and the perception of environmental hazards. *Journal of Health and Human Behaviour* **5**, 144–53.

Vannucci, Pietro 1789. *Discorso istorico–filosofico sopra il tremuoto*. Cesena, Italy.

Varley, A. 1993. *Disasters, development and environment*. New York: John Wiley.

Vay, I. (ed.) 1995. *Il duomo di Venzone: guida breve*. Udine, Italy: Arti Grafiche Friulane.

Ventura, F. 1984. The long-term effects of the 1980 earthquake on the villages of southern Italy. *Disasters* **8**(1), 9–10.

Verluise, P. 1995. *Armenia in crisis: the 1988 earthquake* (translated by L. Chornajian). Detroit, Michigan: Wayne State University Press.

Vinso, J. D. 1977. Financial implications of natural disasters: some preliminary indications. *Mass Emergencies* **2**(4), 205–218.

Vitaliano, D. B. 1973. *Legends of the Earth: their geologic origins*. Bloomington: Indiana University

Press.

Vlasta, M. (ed.) 1996. *Fundamentals of risk analysis and risk management.* Boca Raton, Florida: CRC Press (Lewis Publishers).

Voice, M. E. & F. J. Gauntlett 1984. The 1983 Ash Wednesday fires in Australia. *Monthly Weather Review* **112**(3), 584–90.

Voight, B. 1990. The 1985 Nevado del Ruiz Volcano catastrophe: anatomy and retrospection. *Journal of Volcanology and Geothermal Research* **42**(1/2), 151–88.

Von Buch, Leopold 1824. *Physikalische Beschreibung der Canarischen Inseln.* Berlin.

Wadge, G. (ed.) 1994. *Natural hazards and remote sensing.* London: Royal Society and Royal Academy of Engineering.

Wallace, A. F. C. 1956. *Human behavior during extreme situations.* Disaster Study 1, National Academy of Sciences and National Research Council, Washington DC.

Wang, M. 1981. Consequences of architectural style on earthquake resistance. *Proceedings of the USA–China Joint Workshop on Earthquake Disaster Mitigation*, 150–81. Beijing.

Wang, S-W. & A-C. Zhao 1981. Droughts and floods in China, 1440–1979. In *Climate and history: studies in past climates and their impact on man*, T. M. L. Wigley, M. J. Ingram, G. Farmer (eds), 271–88. Cambridge: Cambridge University Press.

Waterstone, M. (ed.) 1992. *Risk and society: the interaction of science, technology and public policy.* Dordrecht: Kluwer.

Watson Jr, C. C. 1992. GIS aids hurricane planning. *GIS World* (special issue, July), 46–52.

Weiner, D. R. 1988. *Models of nature: ecology, conservation, and cultural revolution in Soviet Russia.* Bloomington: Indiana University Press.

Weiss, T. G. 1997. A research note about military–civilian humanitarianism: more questions than answers. *Disasters* **21**(2), 95–117.

Wen Dazhong 1993. Soil erosion and conservation in China. In *World soil erosion and conservation*, D. Pimental (ed.), 63–85. Cambridge: Cambridge University Press.

Wenger, D. E. & T. F. James 1994. The convergence of volunteers in a consensus crisis: the case of the 1985 Mexico City earthquake. In *Disasters, collective behavior, and social organization*, R. R. Dynes & K. J. Tierney (eds), 229–43. Newark: University of Delaware Press.

Westcoat Jr, J. L., J. U. Chowdhury, D. J. Parker, H. H. Khondker, L. D. James, K. Pitman 1992. *Five views of the flood action plan for Bangladesh.* Working Paper 77, Natural Hazards Research and Applications Information Center, University of Colorado, Boulder.

Wettig, J. & S. Porter 1997. *The Seveso II Directive.* Unpublished document, Directorate-General XI – Environment, Nuclear Safety and Civil Protection, European Commission, Brussels.

White, G. F. 1973. Natural hazards research. In *Directions in geography*, R. J. Chorley (ed.), 193–216. London: Methuen.

—— (ed.) 1974. *Natural hazards: local, national, global.* New York: Oxford University Press.

Whittow, J. 1987. Hazards: adjustment and mitigation. In *Horizons in physical geography*, M. J. Clark, K. J. Gregory, A. M. Gurnell (eds), 307–319. London: Macmillan.

—— (ed.) 1995. Hazards in the built environment. *Built Environment* **21**(2/3), 81–193.

Whyte, A. V. T. 1986. From hazard perception to human ecology. In *Themes from the work of Gilbert F. White: geography, resources and environment* (vol. 2), R. W. Kates & I. Burton (eds), 240–71. Chicago: University of Chicago Press.

Wiggins, J. H. 1996. A reply to "Scientific profiles of the 'Big One'". Disaster Research 181, (Internet bulletin) Natural Hazards Center, University of Colorado, Boulder, Colorado.

Wijkman, A. & L. Timberlake 1984. *Natural disasters: acts of God or acts of man?* Washington DC: Earthscan, International Institute for Environment and Development.

Wilson, A. R. 1991. *Environmental risk: identification and management.* Boca Raton, Florida: CRC Press (Lewis Publishers).

Winchester, P. 1979. Disaster relief operations in Andhra Pradesh, southern India, following the cyclone in November 1977. *Disasters* **3**(2), 173–8.

Winterhalder, B. P. & R. B. Thomas 1978. *Geo-ecology of southern highland Peru: a human adap-*

tation perspective. Occasional Paper 27, Institute of Arctic and Alpine Research, University of Colorado, Boulder.

Wisner, B. 1993. Disaster vulnerability: scale, power and daily life. *GeoJournal* **30**(2), 127–40.

——1995. Bridging "expert" and "local" knowledge for counter-disaster planning in urban South Africa. *GeoJournal* **37**(3), 335–48.

——1998. Marginality and vulnerability: why the homeless don't "count" in disaster preparations. *Applied Geography* **18**(1), 25–33.

Wisner, B., P. O'Keefe, K. Westgate 1977. Global systems and local disasters: the untapped power of peoples' science. *Disasters* **1**(1), 47–58.

Wolensky, R. P. 1979. Toward a broader conceptualization of volunteerism in disaster. *Journal of Voluntary Action Research* **8**(3–4), 33–42.

Wolensky, R. P. & E. J. Miller 1983. The politics of disaster recovery. In *The sociological galaxy: sociology toward the year 2000*, C. E. Babbitt (ed.), 259–70. Harrisburg, Pennsylvania: Beacon's Press for the Pennsylvania Sociological Society.

Wolfe, E. W. 1992. The 1991 eruptions of Mount Pinatubo, Philippines. *Earthquakes and Volcanoes* **23**(1), 5–35.

Wolman, M. G. & J. P. Miller 1960. Magnitude and frequency of forces in geomorphic processes. *Journal of Geology* **68**, 54–74.

World Health Organization 1986. Communicable diseases after natural disasters. *Weekly Epidemiological Record* **11**(14), 79–81.

WRI [World Resources Institute] 1994. *World resources 1994–95: a guide to the global environment.* Oxford: Oxford University Press.

Wyllie Jr, L. A. & J. R. Filson 1989. Armenia earthquake reconnaissance report. *Earthquake Spectra* **5** (special supplement).

Yin, R. K. & G. B. Moore 1985. *The utilization of research: lessons from the natural hazards field.* Washington DC: Cosmos Corporation.

Yoshii, H. 1989. Simulation study of confusion at terminal train stations caused by earthquake warnings. *International Journal of Mass Emergencies and Disasters* **3**(1), 67–86.

Zaman, M. Q. 1989. The social and political context of adjustment to riverbank erosion hazard and population resettlement in Bangladesh. *Human Organization* **48**, 196–205.

Zebrowski Jr, E. 1997. *Perils of a restless planet: scientific perspectives on natural disasters.* Cambridge: Cambridge University Press.

Index

Entries in **bold type** denote illustrations.

INDEX

Peru 186, 220–25
Perugia (Italy) 25–6
Philippines 198–202, **199**, **201**, 208
pillars of modern life **244**
planning for disasters *see* disaster planning
Plinian volcanic eruption 117
Pliny the Elder (*Natural History*) 110, 173
population growth, world 73, 158
positive vulnerability 16
post-traumatic stress disorder 135
poverty 159–62
primary vulnerability 18
"prison of experience" 78, 80
pristine vulnerability 17

quantification 191–2

racism, environmental 187
radial isotropic distribution of impacts **42**, 43, **44**, 45
radical critique of hazards studies 21, 32
Raspe, Rudolph 107
Ray, John 115
reconstruction after disaster 48–50, **183**, 184
recovery from disaster **4**, 48–9
recrimination 134–5
recurrence intervals 8
redundant aid *see* relief of disasters, redundant aid
refugees 211–12
    in Africa **43**
region, types of 28
relief of disasters 47, 69–70, 84–9, 161, 177, 209, 225–6
    appeals and donations 68, 89, 139–41, 147
    as an instrument of conflict 75, 176, 212–13, 216–18; *see also* militarization, warfare
    redundant aid 87
research *see* academic disciplines, academic studies of disaster
resilience to disaster of society 21
resource consumption 70
risk 10–12, 15–16, 192–3, **230**, **234**, **235**, 236, 247
    exposure to 24
    perception of 10–11, 14
    relative 11
road accidents 151
"robber archaeology" 122
Royal Society (London) 115
Rwanda 217

Sacsayhuaman (Peru) **67**
Sakhalin (Russia), 1995 earthquake 89
satellites 154–7, 178
    satellite cycle, the **155**
scenario modelling 168–71
scientific revolutions (Kuhnian paradigms) 108
search and rescue (SAR) 133, 135
secondary vulnerability 19
shelter after disaster 64
Sherman landslide (Alaska) 8
Siena (Italy) 66
slums, favillas, barrios, bidonvilles **92**, 97, 188
social
    change 71–3
    hazards, definition of 9–10, 21
    polarization 71–2
socialization 74
sociocentrism 64
socio-economic development *see* development, socio-economic
sociosphere *see* homosphere, noosphere
Somalia 209–214
"sound-bite culture" 71
space race, the 154–6
spatial models of disaster 40–48
    distribution of impacts in **42**, 43–4, **44**
standardization of tastes 70
Strabo 110
Strategic Defense Initiative 142
structural protection against hazards 24–5
Stukeley, William 112–13
superstition 129
symbiosis 63
syndromes *see* "learned helplessness", "personal invulnerability", "prison of experience"
systems approach 19, 45, 234

taxonomy *see* classification and taxonomy
technocentric view of hazards *see* technocracy
technocracy 31–3, 35, 63, 81, 244–6
"technofix" approach *see* technocracy
technological hazards, definition of 9–10, 21
technology
    military 142, 178
    relationship to hazards 65, 72
telecommunications technology 142, 155
television 137–41, 179–80, 239
theory 227–50

281